MODELING OF HYDRODYNAMICS AND SEDIMENT TRANSPORT IN THE MEKONG DELTA

Thanh Quoc Vo

MODELING OF HYDRODYNAMICS AND SEDIMENT TRANSPORT IN THE MEKONG DELTA

DISSERTATION

Submitted in fulfillment of the requirements of

the Board for Doctorates of Delft University of Technology

and

of the Academic Board of the IHE Delft

Institute for Water Education

for

the Degree of DOCTOR

to be defended in public on

Tuesday, 13 April 2021, at 15.00 hours

in Delft, the Netherlands

by

Thanh Quoc VO

Master of Science in Environmental Management, Can Tho University

born in Bac Lieu, Vietnam

This dissertation has been approved by the
promotor: Prof.dr.ir. J.A. Roelvink and
copromotor: Dr.ir. M. van der Wegen

Composition of the doctoral committee:

Rector Magnificus TU Delft	Chairman
Rector IHE Delft	Vice-Chairman
Prof.dr.ir. J.A. Roelvink	IHE Delft/ TU Delft, promotor
Dr.ir. M. van der Wegen	IHE Delft, copromotor

Independent members:
Prof.dr. A. Ogston	University of Washington, USA
Prof.dr.ir. A.J.F. Hoitink	Wageningen University & Research
Prof.dr.ir. A.J.H.M. Reiniers	TU Delft
Dr. N.H.Trung	Can Tho University, Vietnam
Prof.dr.ir. Z.B. Wang	TU Delft, reserve member

This research was conducted under the auspices of the Graduate School for Socio-Economic and Natural Sciences of the Environment (SENSE)

CRC Press/Balkema is an imprint of the Taylor & Francis Group, an informa business

Published by:
CRC Press/Balkema
Pub.NL@taylorandfrancis.com
www.crcpress.com – www.taylorandfrancis.com
ISBN 978-1-032-04614-3

Dedicated to my beloved parents and family

ACKNOWLEDGMENTS

I would like to take this opportunity to express my heartfelt gratitude to all those who supported me during this study. Completing this dissertation is a great achievement for me. First, I would like to express my deep and sincere gratitude to Prof. Dano Roelvink for giving me the opportunity to do my PhD research under your supervision. I have learnt a lot from you not only knowledge and expertise but also your daily life. I still remember that we had a short talk about an interpolation method for meandering channels. After that, I thought you would never mind about that, but I got your email a day later to solve my problem. This is a critical step for my study.

I would like to thank Assoc.Prof. Mick van der Wegen for your supervision and quick guidance. I was so fortunate to have your quick guidance and response. This speeded up my work. I also thank Johan Reyns for your detailed and interesting lecture before I did my PhD research. Then you supported me for modeling practices. I am also grateful to Dr. Ad van der Spek for taking part in my supervisory team. I gratefully thank the Office of Naval Research for financial supports.

I acknowledge Mr. Herman Kernkamp and Dr. Arthur van Dam for your valuable time to discuss with and guide me on using Delft3D Flexible Mesh. I also thank Dr. Gerald Corzo and Mr. Ander Astudillo (SURFsara) for introducing and allowing me to the SURFsara HPC. It made my modeling simulations possible and saved a lot of time.

PhDers and staff in the group of Coastal Engineering and Port Development are gratefully acknowledged for useful meetings, discussions and beers. I acknowledge Jolanda, Tonneke, Niamh and Martine for their help during my work at IHE Delft.

I thank Can Tho University and my colleagues at College of Environment and Natural Resources and Dragon Institute for their help and encouragement.

Vietnamese people in Delft and the Netherlands are acknowledged for your help and nice conversations which made my life in Delft easy and enjoyable.

I must thank my parents, parents in law, brothers and sister for their unconditional support. My heartful thank goes to my wife and my daughter for their love and support.

As a final word, I would like to thank each and every individual who have been supported and encouraged me to achieve my goal and complete my dissertation successfully.

Vo Quoc Thanh

SUMMARY

Deltas are low-lying plains which are formed when river sediments deposit in coastal environments. Deltas are nutrient-rich, and productive ecological and agricultural areas with high socio-economic importance. Globally, deltas are home to about 500 million people and are considerably modified by human activities. In addition, they are vulnerable to climate change and natural hazards like changing river flow and sediment supply, coastal flooding by storminess or sea level rise. To encourage better delta management and planning, it is of utmost importance to understand existing delta sediment dynamics.

The objective of this study is to investigate the prevailing sediment dynamics and the sediment budget in the Mekong Delta by using a process-based model. Understanding sediment dynamics for the Mekong Delta requires high resolution analysis and detailed data, which is a challenge for managers and scientists. This study introduces such an approach and focuses on modeling the entire system with a process-based approach, Delft3D-4 and Delft3D Flexible Mesh (DFM). The first model is used to explore sediment dynamics at the coastal zone. The latter model allows straightforward coupling of 1D and 2D grids, making it suitable for analysing the complex river and canal network of the Mekong Delta.

This study starts by generating trustworthy bathymetries based on limited data availability. It describes a new interpolation method for reproducing the main meandering channel topographies of the Mekong River. The reproduced topographies are validated against high resolution measured data. The proposed method is capable of reproducing the thalweg accurately.

Next, this study describes the development of a Delft3D Mekong Delta model. The model is validated for hydrodynamics and sediment dynamics data for several years and focuses on describing near shore sediment dynamics. The model shows that sediment transport changes in the Mekong Delta are strongly modulated by seasonally varying river discharges and monsoons. The nearshore suspended sediment concentration (SSC) is significantly decreased due to a lack of wave-induced stirring when there is no monsoon. 3D Gravitational circulation effects limit the SSC field from expanding seaward in case of high river flow. In addition, the bed composition has an important role in reproducing sediment fluxes which were considerably decreased when a sandy bed layer is included. This happens due to effects of the initially mostly sandy mixing layer, where resuspension of the mud is proportional to the fraction of mud present. It takes time for an equilibrium bed composition to develop. Seasonally, the sediment volumes deposited in the river mouths increase regularly during the high flow season. During October they remain more or less constant and then, as wave action increases and discharges decrease, the deposited material is resuspended and transported southward along the coast.

The DFM model explores the hydrodynamics and sediment dynamics in the fluvial reach of the Mekong River including the anthropogenic effect of dyke construction. After an extremely high flood in 2000 which caused huge damages, a dyke system has been built to protect agriculture in the Vietnamese Mekong Delta (VMD). These structures change hydrodynamic characteristics on floodplains by avoiding floodwaters coming into the floodplains. The DFM model shows that the high dykes slightly change hydrodynamics in the VMD downstream. These structures increase daily mean water levels and tidal amplitudes along the mainstreams. Interestingly, the floodplains protected by high dykes in Long Xuyen Quadrangle and Plain of Reeds influence water regimes not only on the directly linked Mekong branch, but also on other branches.

Based on the validated hydrodynamic model, the model is validated against sediment data and used to derive a sediment budget for the Mekong Delta. For the first time, this study has computed sediment dynamics over the entire Mekong Delta, considering riverbed sediment exchange. The model suggests that the Mekong Delta receives ~99 Mt/year sediment from the Mekong River This is much lower than the common estimate of 160 Mt/year. Only about 23% of the modelled total sediment load at Kratie is exported to the sea. The remaining portion is trapped in the rivers and floodplains of the Mekong Delta. Located between Kratie and the entrance of the Mekong Delta, the Tonle Sap Lake receives Mekong River flow at increasing flow rates seasonally and returns flow when Mekong River flow rates decay. As a result Tonle Sap Lake traps approximately 3.9 Mt/year of sediments and explains the hysteresis relationship between water discharges and SSC at downstream stations. The VMD receives an amount of 79.1 Mt/year (~80 % of the total sediment supply at Kratie) through the Song Tien, the Song Hau and overflows. The model results suggest that the Mekong mainstream riverbed erodes in Cambodia and accretes in Vietnam.

The results of this study advance understanding of sediment dynamics and sediment budget in the Mekong Delta. The model developed is an efficient tool in order to support delta management and planning. The validated model can be used in future studies to explore impact of climate change and human interference in the Mekong Delta.

SAMENVATTING

Deltas zijn laagvlaktes gevormd door het neerslaan van door rivier aangevoerd sediment langs de kust. Deltas vormen vaak productieve landbouw gebieden van aanzienlijk socio-economische belang. Wereldwijd bieden deltas een leefomgeving voor ongeveer 500 miljoen mensen en worden mede gevormd voor en door menselijke aanwezigheid en activiteit. Daarnaast zijn deltas kwetsbaar voor klimaat verandering en natuur geweld zoals veranderende rivier stroming en sediment toevoer, overstroming van kustgebieden door stormen of zeespiegelstijging. Het is van essentieel belang om de bestaande dynamica van deltas te doorgronden om daarmee duurzaam beheer mogelijk te maken.

Het doel van deze studie is om de heersende sediment dynamica en het sediment budget in the Mekong Delta te onderzoeken middels een proces-gebaseerd model. Het begrijpen van de sediment dynamica van de Mekong Delta vereist een hoge resolutie analyse op basis van gedetailleerde data, wat een uitdaging vormt voor managers en wetenschappers. Deze studie levert zo'n analyse en richt zich op het modeleren van het gehele system met een proces-gebaseerd model, namelijk Delft3D-4 en Delft3D Flexible Mesh (DFM). Het eerste model wordt gebruikt om de sediment dynamica te bestuderen in de kustzone. Het tweede model staat efficiënt koppelen van 1D netwerken en 2D roosters toe waardoor het geschikt is om het complexe rivier en kanaal netwerk van de Mekong Delta te modeleren.

Deze studie start met het genereren van betrouwbare bathymetrieën op basis van beperkte bodem data beschikbaarheid. Een nieuwe interpolatiemethode wordt beschreven om bathymetrieën van de meanderende hoofdgeul van de Mekong Rivier te reproduceren. Vervolgens worden deze bathymetrieën gevalideerd ten opzichte van hoge resolutie data. De voorgestelde methode blijkt in staat om de talweg accuraat te beschrijven.

Vervolgens beschrijft deze studie de ontwikkeling van het Delft3D Mekong Delta model. Het model is gevalideerd met hydrodynamische en sediment-dynamische data over verschillende jaren en richt zich op het beschrijven van kust gerelateerde sediment dynamica. Het model laat zien dat veranderingen van sediment transport in de Mekong Delta sterk beïnvloed worden door seizoen variërende rivier afvoer en moessons. De gesuspendeerde sediment concentratie (SSC) aan de kust vermindert significant als er geen moesson is en dus geen golf geïnduceerde suspensie van sediment. 3D zwaartekracht gedreven circulaire stroming beperkt zeewaartse uitbreiding van het SSC veld in het geval van hoge rivier afvoer. Daarnaast speelt de bodem samenstelling een belangrijke rol in het reproduceren van sediment fluxen die sterk verminderen in het geval van een zandige bodem. Dit gebeurt doordat slib re-suspensie vanuit de bodem proportioneel is aan de fractie aanwezige slib als de bodem initieel voornamelijk bestaat uit zand. Een bodem heeft tijd nodig om zich te ontwikkelen tot een bodem met een evenwicht samenstelling van sediment fracties. Op een tijdschaal van seizoenen nemen de sediment volumes die neerslaan in de riviermonding toe gedurende hoge afvoer. In oktober blijven de volumes min of meer constant en vervolgens, wanneer golf actie en rivier afvoer toenemen, wordt het neergeslagen sediment opnieuw in suspensie gebracht en zuidwaarts langs de kust getransporteerd.

Het DFM model verkent de hydrodynamica en sediment dynamica in het meer fluviale deel van de Mekong Delta, inclusief de antropogene effecten van dijk constructie. Na een extreme hoge rivier afvoer, welke grote schade veroorzaakte in 2000, is een dijksysteem ontwikkelt om de landbouw in de Vietnamese Mekong Delta (VMD) veilig te stellen. Deze dijk constructie heeft de hydrodynamica op de uiterwaarden veranderd door overstroming van de uiterwaarden te beperken. Het DFM model laat zien dat de hoge dijken de benedenstroomse hydrodynamica enigszins beïnvloeden. De dijken verhogen de dagelijkse gemiddelde waterstanden en getij amplituden langs de voornaamste rivier takken. Interessant genoeg, beïnvloeden de door dijken beschermde uiterwaarden in de Long Xuyen rechthoek en de Vlakte van het Riet niet alleen het stromingsregime van de direct nabij gelegen rivier tak, maar ook van de andere takken van de Mekong rivier.

Aan het eerder gevalideerde hydrodynamische model is een sediment transport module toegevoegd en gevalideerd met sediment data om een sediment budget voor de Mekong Delta af te leiden. Voor het eerst heeft deze studie sediment dynamica berekend over de hele Mekong Delta met in acht name van sediment uitwisseling met de bodem. Het model suggereert dat de Mekong Delta ongeveer 99 Mt sediment per jaar ontvangt van de Mekong rivier. Dit is veel minder dan de gewoonlijk aangenomen 160 Mt per jaar. Slechts ongeveer 23% van de totale sediment toevoer vanaf Kratie (de stroomopwaartse randvoorwaarde van het model) wordt geëxporteerd naar de zee. Het resterende deel wordt afgezet in de rivieren en uiterwaarden van de Mekong Delta. Gelegen tussen Kratie en de ingang van de Mekong Delta, ontvangt het Tonle Sap Lake afvoer van de Mekong rivier tijdens oplopende rivier afvoer en draagt het aan de Mekong rivier afvoer bij tijdens verminderende rivier afvoer. Daardoor vangt het Tonle Sap Lake ongeveer 3.9 Mt sediment per jaar in. Dit proces verklaart ook de vergroting van het hysteresis effect tussen rivier afvoer en SSC bij stroomafwaarts gelegen stations. De VMD ontvangt 79.1 Mt sediment per jaar (ongeveer 80% van de totale sediment toevoer vanaf Kratie) door de Song Tien, de Song Hau, en de uiterwaarden. De model resultaten suggereren dat de bodem van de voornaamste Mekong rivier tak erodeert in Cambodia en verhoogt in Vietnam.

Deze studie draagt bij aan een beter begrip van de sediment dynamica en sediment budget van de Mekong Delta. Het ontwikkelde model is een efficiënt middel om delta beheer en planning te ondersteunen. Het gevalideerde model kan gebruikt worden om de impact van klimaat verandering en menselijk ingrijpen in de Mekong Delta te verkennen.

CONTENTS

1
INTRODUCTION

Abstract

This chapter presents the research framework of this study. First, it describes an overview of hydrodynamics and sediment transport in deltas, focusing on the Mekong Delta. This delta is facing several challenges which significantly change hydrodynamic and morphodynamic behaviours of the Mekong Delta. The research problems are identified and important roles of numerical models are illustrated. Second, the objective of this study are shown and the research questions are formulated. Finally, the structure of this dissertation is shown.

1.1 BACKGROUND

Deltas are important to human activities because they are home to about 500 million people (Syvitski et al., 2009). A river delta is low-lying plain where river meet the sea so it has both riverine and marine characteristics (Nguyen Anh Duc, 2008). Therefore, deltas are the most productive ecological areas and have abundant wildlife and high biodiversity, having saline and fresh water, riverine and marine sediments. Typically, estuaries are defined as a transition zone between river and ocean environments. Deltas are not only great areas in biodiversity but also in human settlements of approximately 60% of the world's population. Of the 32 largest cities, 22 are located on estuaries (Karamouz et al., 2013). Consequently, deltas have been influenced by human activities such as land-use changes, port development, land reclamation, diking, damming of channels and dredging, and sand mining. These anthropogenic factors also contribute to natural fluctuations and comprise modifications of hydrodynamics (Marineau and Wright, 2014), sediments discharge (Manh et al., 2015; Renaud et al., 2013) and morphology (Dissanayake and Wurpts, 2013). In addition, IPCC (2007) documented that deltas are highly vulnerable areas due to climate change. They are continuously impacted by climate change, including sea level rise and other natural hazards (Renaud et al., 2013). According to recent assessments, 40 deltas in the world are projected to be at risk in terms of coastal erosion and sinking due to decrease of sediment and sea level rise. Among these, the Ganges-Brahmaputra delta in Bangladesh, the Vietnamese Mekong Delta (VMD) and the Nile delta in Egypt are highlighted as extremely vulnerable deltas by predicted sea level rise up to 2050 (IPCC, 2007). Because of this, deltas are highly dynamic areas and their environmental conditions can be significantly changed. They are low land areas (about 26,000 square kilometers below local mean sea level in thirty-three major deltas around the world) so they may be very vulnerable to sea level rise (Overeem and Syvitski, 2009). However, climate change and other natural changes are not the only key drivers controlling deltas; another more important factor is human activity. Rivers have been used as a fresh water resource for agricultural irrigation and hydropower dams, leading to changes in river morphology and hydrodynamics. Globally, large reservoirs are filled by over 40% of river discharge, trapping 26% of river sediment to the deltas and coastal zones (Overeem et al., 2013).

The case study covers areas of hydrological sub-zone 5 and 6 (Figure 1.1). The Mekong River is one of the largest rivers around the world, ranking eighth in mean annual water discharge (14,500 m^3/s at the river mouth), tenth in suspended sediment discharge (110 million tons per year) and twelfth in river length (4880 km) (Gupta et al., 2006; Gupta and Liew, 2007; Milliman and Farnsworth, 2011; MRC, 2011). There are about 60 million people living in the Lower Mekong Basin (LMB), whose livelihood depend on natural resources of the Mekong River (Kummu and Varis, 2007). As a consequence of development, numbers of hydropower dams and reservoirs have been building in the

Mekong Basin. Particularly, the Manwan Dam decreased dramatically of 56% (about 40 million tons annually) in total suspended solids (including sediments) after its closure in 1993 (Kummu and Varis, 2007). The dams and reservoirs may cause changes of LMB's hydrology and river morphology. Moreover, climate change and sea level rise also change the hydrological regime and sediment transport in the Mekong Delta.

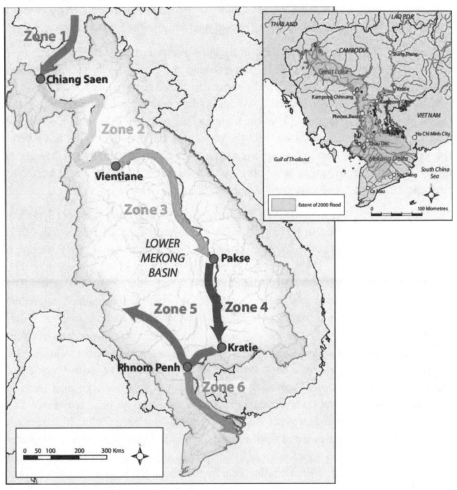

Figure 1.1. Hydrological sub-region of LMB (MRC, 2009b).

The Mekong Delta is important to local livelihoods and food security. It home to about 18 million people in Cambodia and Vietnam (Tran, 2016). The VMD has an area of ~39,700 km^2 and land-use for agriculture occupies ~60% of this area (Vo, 2012). Thus it contributes to the majority of rice export in Vietnam. Cultivating agricultural crops in the VMD significantly depend on water sources from the Mekong River. It supplies ~416 km³ of yearly water volume and delivers ~73 Mt per year of sediment at Kratie (Koehnken, 2012; Thanh et al., 2020a). Therefore, the changes of hydrodynamics in the Mekong River will considerably influence hydrodynamics and sediment transport the Mekong Delta. In particular, changes of sediment transport cause sedimentation and morphology changes in the VMD. Thus it is necessary to understand mechanism of sediment transport.

Recently, the Mekong has become one of the most active regions for hydropower development in the world. MRC (2011) reported that there were totally 136 existing and planned dams in the LMB in which hydropower dams have been mainly built in Laos and Cambodia. Unfortunately, Laos and Cambodia mainly contribute water discharge to the Lower Mekong River and these dams are trapping sediment that is the main resource for delta development. In combination with the context of sea level rise, VMD has several challenges in terms of hydrodynamics and morphology changes. Therefore, it is essential to investigate hydrodynamics and morphodynamics in the delta in the contexts of anthropogenic impacts, climate change and sea level rise.

There are a considerable number of studies focusing on investigating sediment dynamics in the VMD (Heege et al., 2014; Hung et al., 2014b; Loisel et al., 2014; Manh et al., 2013, 2014; Marchesiello et al., 2019; McLachlan et al., 2017; Nowacki et al., 2015; Stephens et al., 2017; Thanh et al., 2017; Tu et al., 2019; Wolanski et al., 1996; Xing et al., 2017). Depending on the objectives, these studies used different approaches and scales. The common approaches used are in-situ measurement, remote sensing and numerical modeling. Each technique has advantages and disadvantages. Remote sensing allows coverage of large areas. It enables regional observations of suspended sediment concentration, but it is impossible to investigate mechanisms of sediment transport and deposition. This problem can be solved by in-situ measurement or/and numerical modeling. Nevertheless, in-situ measurement is difficult to apply in large areas because it is time consuming and costly. Numerical modeling seems to be a reasonable and efficient technique for a large area such as the whole Mekong Delta. In numerical modeling, the spatial scales are in a reverse ratio to model details. For instance, Manh et al. (2014) used a 1D model for the entire Mekong Delta to investigate sediment transport and sediment deposition. With a smaller scale for the Mekong tidal river and subaqueous delta, Thanh et al. (2017) and Tu et al. (2019) used a process-based model (Delft3D) to investigate sediment dynamics and morphological changes. Actually, these modeling approaches can be combined as an operational model train which consists of a large scale and coarse model and a small scale and detailed model (Thanh et al., 2017). Outputs of the large and coarse model are defined as boundary conditions of the small and detailed model. This approach may increase uncertainty of the modelled results due to coupling computation. Another limitation is that it significantly increases the number of simulations, leading to difficulties for model calibration. Consequently, it is envisaged that creating a single modeling domain for the entire Mekong Delta and its shelf could result in more accurate results.

1.2 OBJECTIVE AND RESEARCH QUESTIONS

The objective of this study is to investigate the sediment dynamics and sediment budget in the Mekong Delta.

To achieve the objective of this research, four research questions are formulated. These research questions are capable of guiding this research implementation. These questions are the following.

1. *How can a 2D/3D model be applied in cases of limited topography data?*

2. *What is the role of coastal processes in sediment modeling for the Mekong Delta?*

3. *How does the delta-based water infrastructure influence hydrodynamics in the Mekong Delta?*

4. *What are prevailing sediment dynamics and sediment budget in the Mekong Delta?*

1.3 METHODOLOGY

In order to answer the research questions, hydrodynamics and sediment dynamics of the Mekong River are investigated. Recently, there are a number of studies focusing on these topics (e.g. Dinh et al., 2012; Dutta et al., 2007; Fujii et al., 2003; Heege et al., 2014; Hung et al., 2014b, 2014a; Kite, 2001; Le et al., 2007, 2008; Manh et al., 2015, 2014; Nguyen et al., 2008; Nguyen Van Manh, 2014; Nowacki et al., 2015; Thanh et al., 2017, 2020a; Tri et al., 2012; Tu et al., 2019; Unverricht et al., 2013; Van et al., 2012; Vinh et al., 2016; Wassmann et al., 2004; Xue et al., 2012). Based on their approaches and methods applied, these studies can be grouped, as field measurements, remote sensing and numerical modeling. In general, a numerical model is an efficient tool in order to understand sediment dynamics in a complex system as the Mekong Delta and the others could be importantly supportive.

This study applies a process-based model which solves the shallow water equations based on the finite volume numerical method (Kernkamp et al., 2011). The Mekong Delta consists of a dense river network, with high variability of channel widths, particularly in the VMD. The river network encompasses natural rivers, man-made canals and floodplains and is a result of water infrastructure development from 1819 onwards (Hung, 2011). Thus a pure 2D model for the entire Mekong Delta is inefficient since it increases the number of computational nodes. Besides, 1D models are efficient in large areas such as the Mekong Delta system, but they are not able to consider the river-sea interaction. The river-sea interaction is very important in sediment transport modeling, as was shown in recent studies (Thanh et al., 2017; Tu et al., 2019; Xing et al., 2017). Therefore, a combination of these two seems a reasonable solution. A hybrid modeling grid which includes 2D cells and 1D elements, is one of optimal and efficient approaches for the Mekong Delta. Moreover, available data of bathymetry of the Mekong River are limited and coarse. This needs a higher resolution of bathymetry data for the 2D cells, so this study introduces a spatial interpolation method for meandering channels, based on the channel-fitted coordinates.

1.4 THESIS OUTLINE

This thesis starts with an introduction to this study which is presented in the Chapter 1. Following the introduction, Chapter 2 introduces an efficient method for spatial interpolating topography of a meandering channel. An important step of this method is removal of anisotropic effects. The method suggested was validated with the measured topographic data of a Mekong River branch (Cua Tieu). Results of validation suggested that the method is appropriate for sparse and limited data areas. It was also applied to reproduce topography of the Mekong River which was used for modeling in this study. Chapter 3 describes a multiscale modeling approach for the Mekong Delta. This approach includes a large-scale and coarse model and a small-scale and detailed model. Outputs of the large-scale model were defined as boundary conditions for the small-scale model. The model train reduces computational work. Chapter 4 explores the impact of delta-based water infrastructure (high dykes) on downstream hydrodynamics. The impact was quantified by using a 1D-2D coupled model. This Chapter also presents scenarios of high dyke development and their effects on water levels and tidal propagation in the downstream. Subsequently, Chapter 5 investigates the sediment dynamics and sediment budget in the entire Mekong Delta using the 1D-2D coupled model. The large-scale model allows consideration of fluvial and coastal processes. Moreover, the hybrid modeling grid includes 2D cells for the Mekong River mainstream, floodplains and shelf and 1D elements for the primary and secondary canals/rivers in the Mekong Delta. The 1D-2D coupling enables optimal computational work. Finally, Chapter 6 summarizes this research and answers the research questions. It also presents recommendations for further studies.

2

SPATIAL TOPOGRAPHIC INTERPOLATION FOR MEANDERING CHANNELS[1]

Abstract

Bathymetric data plays a major role in obtaining accurate results in hydrodynamic modeling of rivers, estuaries and coasts. Bathymetries are commonly generated by spatial interpolation methods of data on a model grid. Sparse and limited data will impact the quality of the interpolated bathymetry. This study proposes an efficient spatial interpolation framework for producing a channel bathymetry from sparse, cross-sectional data. The proposed approach consists of three steps: (1) anisotropic bed topography data locations transformed to an orthogonal and smooth grid coordinate system that is aligned with its river banks and thalweg; (2) sample data are linearly interpolated to generate river bathymetry; and (3) the generated river bathymetry is converted into its original coordinates. The proposed approach was validated with a high spatial resolution topography of the Tieu estuarine branch. In addition, the proposed approach is compared to other spatial interpolation methods such as ordinary kriging, inverse distance weighting, and kriging with external drift. The proposed approach gives a nearly unbiased topography and a strongly reduced RMSE compared to the other methods. In addition, it accurately reproduces the thalweg. The proposed approach appears to be efficiently applicable for regions with sparse cross-sections. Moreover, river topography generated by the proposed approach is smooth including important morphologic features, making it suitable for two- and three-dimensional hydrodynamic modeling.

---

[1] This chapter is based on:

Thanh, V. Q., Roelvink, D., van der Wegen, M., Tu, L. X., Reyns, J. and Linh, V. T. P.: Spatial topographic interpolation for meandering channels, J. Waterw. Port, Coastal, Ocean Eng., 146(5), 04020024, 2020.

2.1 INTRODUCTION

Topography of rivers, estuaries and coasts plays a crucial role in investigating hydrodynamic processes, water-related contaminant transport and morphological changes because it strongly influences modelled results (Conner and Tonina, 2014). Therefore, topographies should be accurate and detailed. Acquiring detailed bathymetry data is difficult and costly. River bathymetry can be generated by field surveys and remote sensing images (Conner and Tonina, 2014; Dilbone et al., 2018; Legleiter, 2013). However, remote sensing seems only applicable in clear-flowing and gravel-bed rivers/channels. In cases of muddy channels with high suspended sediment concentration, remote sensing cannot predict accurate river bathymetry due to limited signal penetration. Details and accuracy of river bathymetries by field survey, depend on density of sampling points acquired. If the sampling points are in low density, they need an interpolating method. River bathymetry interpolation methods are manifold, like original spatial or controlling-directional interpolations (Bailly du Bois, 2011; Carter and Shankar, 1997; Caviedes-Voullième et al., 2014; Chen and Liu, 2017; Conner and Tonina, 2014; Goff and Nordfjord, 2004; Hilton et al., 2019; Lai et al., 2018; Legleiter and Kyriakidis, 2008; Lin and Chen, 2004; Merwade, 2009; Merwade et al., 2008; Sear and Milne, 2000; Zhang et al., 2016).

River bathymetries in 1D models are represented by cross-sectional data. Distances between cross-sections are usually long. For instance, cross-sections of the 1D ISIS model for the Mekong Delta have distance intervals ranging from 500 to 3,000 m. However, a 2D model of the Mekong River has a grid resolution of 300-600 m (Thanh et al., 2017, 2020a). Therefore, the cross-section data is insufficient for the 2D model and require a spatial interpolation method. There are a number of spatial interpolation methods particularly suitable for river bathymetries, such as linear, inverse distance weighting (IDW) and some kriging methods. These methods are efficiently applied for isotropic data. However, river bathymetry data is strongly influenced by river flows, so it has a certain longitudinal trend. If applications of these interpolation methods do not take into account known spatial trends, they may generate inaccurate river topography (Merwade, 2009). Therefore, eliminating longitudinal trends of river bathymetry before applying interpolation methods would give a better predictions. The spatial trends of bathymetric data can be excluded by some approaches, such as converting the data into river-aligned coordinates or forcing metrics. Rivest et al. (2008) conducted a study to have better predictions by converting testing data from the Cartesian grid into the natural coordinates of flow. This improves accuracy of kriging methods. Legleiter and Kyriakidis (2008) introduced a geostatistical framework to predict river topography. The framework includes steps of (1) transformation of data into channel-centred coordinates and (2) estimating river bed elevations. Some kriging methods were applied to estimate the bed elevations, including universal kriging, ordinary kriging with breaklines, kriging with an external drift in which a simple trend is considered based on relationship between planform and cross-section asymmetry. Merwade (2009) applied a similar framework. First, locations of bathymetric sample points are converted to *sn* coordinates based on center lines. Then the interpolation methods of IDW, regularized spline, spline with tension, topogrid, natural neighbour (NN), ordinary kriging (OK) and OK with anisotropy were applied for six river reaches. They conclude that it is difficult to determine the best interpolation method due to different sampling densities and distribution. The best interpolators are changed depending on characteristics of sampling data (different river reaches). Zhang et al. (2016) developed an interpolation method, called shortest temporal distance. This method is to reduce effects of data anisotropy by using temporal distances metrics. Their method is validated and compared to UK and IDW. Besides, Chen and Liu (2017) compared the three methods of linear interpolation, IDW and NN in resampling cross-sections. Their finding is that the linear interpolation is a good method which is able to maintain morphologic features in meandering rivers. In summary, a general and common approach for generating river bathymetry is excluding effects of data anisotropy and then applying a spatial interpolation method. A common way to diminish anisotropic effects is converting to centred-line coordinates (Goff and Nordfjord, 2004; Legleiter and Kyriakidis, 2008; Merwade, 2009).

The accuracy of an interpolated bathymetry highly depends on density and spatial distribution of the sampling data (Merwade, 2009). Studies by Legleiter and Kyriakidis (2008) and Zhang et al. (2016) are based on high resolution data (7 m and 50 m, respectively) while Merwade (2009) used separated data occupying 70% of the total samples for interpolation. However, in cases with limited and sparse data such as the Mekong River, Vietnam this interpolation framework should be modified. The most common 2D river topography of the Mekong River, Vietnam is derived from cross-sectional data from 1D hydrodynamic models (Dung et al., 2011; Manh et al., 2014; Tran et al., 2018; Triet et al., 2017; Van et al., 2012; Wassmann et al., 2004). Cross-sectional data from these 1D models is sparse, with cross-section spacing of 500 - 3,000 m.

This study aims to propose an efficient spatial interpolation framework, is called anisotropy-removed interpolation method (AR), for generating river and estuarine bed topography from sparse cross-sections. The framework was implemented by three steps: (1) anisotropic bed topography data locations transformed to a channel-fitted coordinate based on river banks and thalweg; (2) sample data are linearly interpolated to generate bed topography; and (3) the generated river bathymetry is converted into its original coordinates. The testing data is the river topography of the Tieu estuarine branch in the Mekong Delta. The cross-sections used for interpolation are extracted from a high spatial resolution of around 50 m. Distances between these cross-sections range from 500 to 2,000 m. A fine and smooth 2D grid that aligns the river reach is generated based on the river banks and the thalweg instead of the centre line between two banks. This step is to

9

exclude anisotropic effects and adds to generate a continuous meandering thalweg. After that the linear interpolation method is used to produce a smooth river bed surface. Performance of this framework is validated and presented by statistical indices of coefficient of correlation, mean error, and root mean square error in comparison with some commonly used methods of IDW, OK, and kriging with external drift (KED).

2.2 METHOD

2.2.1 Data

The Tieu branch is one of the main Mekong River's branches. . The Tieu branch is the smallest branch of the Mekong River in terms of river width and cross-sectional area. Its width and cross-sectional area are 1,100 m and 7,100 m2, respectively (Nguyen Anh Duc, 2008). The river length that contains topographic samples is approximately 15 km (Figure 2.1). The river width at its mouth is 1,100 m and dramatically decreases landward to around 400 m at the west boundary of the topography.

The topographic samples are collected by the Southern Institute of Water Resources Research, Vietnam in 2010 by an ODOM HYDROTRAC echosounder. Figure 2.2 presents histogram of sample elevation which was analysed by 2056 samples. The mean and standard deviation of sample elevation are -6.9 m and 2.9 m respectively. These samples have a relatively uniform distribution in space (Figure 2.1). Due to the fact that the objective of this study is to propose an interpolation approach for sparse and discrete cross-sections, several river cross-sections are extracted at every 500-2,000 m interval. These intervals are relatively similar to distances between cross-sections in 1D hydrodynamic models for the Mekong Delta, e.g. ISIS (Van et al., 2012). The cross-section spacing is still smaller than six times river width. If it is higher, the interpolated bathymetry would miss the main morphologic features (Conner and Tonina, 2014). Distances between cross-sections extracted depend on the river meandering. The river segments bounded by the extracted cross-sections are as straight as possible. As a result, there are 15 extracted cross-sections, with a number of 186 samples. This number of samples is of about 9% of the total topographic samples. The remaining samples are used to validate the proposed interpolating approach.

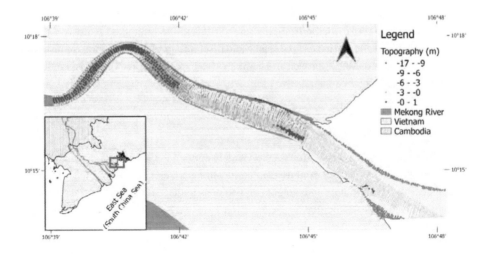

Figure 2.1. Location of the Tieu branch and its bed topography collected by the Southern Institute of Water Resources Research, Vietnam in 2010.

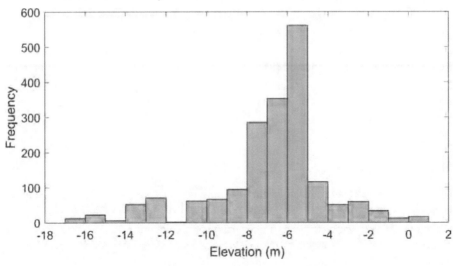

Figure 2.2. Histogram of sample of the Tieu branch.

2.2.2 Selected methods for river bed topography interpolation

There is an increasing amount of interpolation methods applied for river topography estimation (e.g. Bailly du Bois, 2011; Carter and Shankar, 1997; Caviedes-Voullième et

11

al., 2014; Curtarelli et al., 2015; Merwade, 2009; Merwade et al., 2008; Zhang et al., 2016). The most popular methods applied are IDW and OK methods. However, these methods may be inefficient in cases of sparse and discrete samples. Therefore, we propose an alternative interpolating approach and compare this to most applied methods, i.e. IDW, OK and KED described below.

Inverse Distance Weighting

IDW is a deterministic interpolation method widely applied in GIS software packages (Li and Heap, 2011). The values of non-sampled locations are estimated from the values of several nearby sample points and weights of distances to these points. The values of non-sampled points ($\hat{z}$) are computed by the equation 1.

$$\hat{z} = \frac{\sum_{i=1}^{n}(\frac{z_i}{d_i^p})}{\sum_{i=1}^{n}(\frac{1}{d_i^p})} \ (1)$$

where n is the number of sample points considered; z_i is the value at the i location; d_i is the distance between the estimated point and sample point i and p is the exponent of a power function. This formula includes two external factors influencing the estimated value, namely the density of the considered sample points and the exponent. In order to apply IDW to create topography, the important factor is the exponent p which accounts for the importance of distance of different sample points. A higher exponent value puts less importance to longer distance sample points. Commonly used values are p = 1 or 2. For the samples used in this study, the exponent value was set to two in the case with a low sampling density.

Ordinary Kriging

OK is the most commonly used kriging method. OK estimates values at non-sampled locations based on the spatial structure of sample points' attribution. Similar to IDW, the attribution of non-sampled points is estimated by neighbour sample points, but with different weights. The weights in the kriging interpolation are statistically specified by the semivariogram. The weight given to each observation depends on the degree of spatial correlation. The semivariogram depicts the spatial autocorrelation of the sample points in accordance to their distances and it is calculated by the following equation.

$$\gamma\ (h) = \frac{1}{2n} \sum_{k=0}^{n}(z_i - z_{i+h})^2 \ (2)$$

where z_i is the depth value at the sample i; z_{i+h} is the depth value of a neighbour sample, with distance h from the i location; and n is the number of sample pairs.

All pairs of the sample locations are plotted and a fitting model is used to present the pattern of relationship. For selecting a fitting model, two components need to be considered, encompassing the spatial autocorrelation and the semivariogram model. The former is described by the sample data through certain characteristics which are the range, the sill and the nugget. Semivariogram models are usually described by Exponential, Spherical, Gaussian, Matern and Linear functions of which the most common type is the Spherical function. In addition, the semivariograms can be defined for a specific direction. However, in the case of meandering, it is really difficult to define a specific direction so the direction is not taken into account in this study.

Kriging with External Drift

Obviously, a channel has a clear trend with aligning river banks following the flow direction. Therefore, to consider effects of the channel direction, an external drift is included in kriging interpolation, called KED. This dramatically reduces anisotropy effects when taking into account the thalweg of a channel. Distance-to-thalweg is a crucial factor to eliminate effects of anisotropy in interpolating river bathymetry (Wille, 2013). In this case, the unknown samples are predicted as in kriging, but with a different covariance matrix of residuals (Webster and Oliver, 2007). In this study, these interpolation methods were implemented in the free software environment of R, with the *gstat* geostatistics packages, introduced by Pebesma (2004).

Anisotropy-removed interpolation method

There are numerous studies using 1D models , e.g. for the Mekong Delta (Manh et al., 2014; Tran et al., 2018; Triet et al., 2017; Van et al., 2012; Wassmann et al., 2004). The data representing the river topography consist of cross-sections. In order to derive 2D river topography from these cross-sections, an interpolation method is needed.

Isotropic interpolation methods are not suitable because of the anisotropic channel morphology. Merwade et al. (2008) applied isotropic interpolation methods in a transformed coordinate based on the central line while this study used a transformed coordinate based on the river banks and the thalweg. This study introduces an interpolation method specifically for this type of river topography data. The interpolation process is illustrated by the following three steps and is implemented in Matlab.

- **Step 1**: The river segment is presented by an orthogonal and smooth curvilinear grid. The grid is generated based on the river banks and the thalweg and it is processed to be orthogonal and smooth in order to accurately represent the river wetted areas. Figure 2.3 represents the referenced grid and the samples of cross-sections. The resolution of the grid is fine enough to capture all the samples. This means that the grid sizes should be smaller than the distance between the two closest samples.

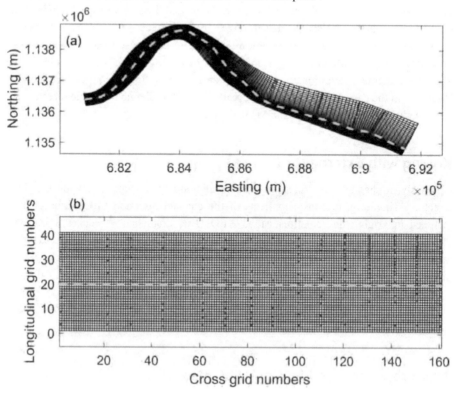

Figure 2.3. The grid used in interpolation (a) and the straightened grid transformed into curvilinear coordinates (b). The green dashed line is the thalweg line and the blue dots are sampling points for interpolation.

- **Step 2**: The cross-section data in a Cartesian coordination system is transformed into a curvilinear coordinate. Specifically, the grid is straightened in the horizontal and vertical coordinates, representing the longitudinal and cross directions of the river reach. The curvilinear grid cells are converted to rectangular grid cells (Figure 2.3.). Simultaneously, the samples are also located in the transformed coordinates. This step is to reduce effects of anisotropy caused by river flows. In the transformed coordinates, the river is presented as a rectangular reach. This general approach is also applied by Merwade (2009). However, in case of meandering or braided channels, this approach may be insufficient to reproduce continuous thalweg lines. It leads to misleading predictions of the thalweg that generates unrealistic ripples between cross-sections. Therefore, we take the thalweg as a reference for coordinate transformation. The thalweg line, which is the deepest path along the channel, is easily generated based on high resolution bathymetry. However, it is difficult to identify the thalweg correctly in sparse data channels. It is theoretically defined by the horizontal shapes of channels. For instance, the thalweg would be near the outside bank of bends (Loucks, 2008). In these channels with cross-section data, we suggest that the thalweg between their adjacent cross-sections regularly moves along the channels between the thalweg locations on these cross-sections.

-    **Step 3**: In the curvilinear coordinate system, the grid corners (unsampled points) are estimated by any spatial interpolation. To interpolate the data between two cross-sections, it is suggested to apply a linear interpolation method along the river (Deltares, 2018). Therefore, in this study, in order to maintain continuous wetted areas of river's interpolated cross-sections, we used a linear interpolation based on a triangulated irregular network (TIN) which derives a bivariate function for each triangle to estimate unsampled points' depths (Mitas and Mitasova, 2005). This accepts the assumption that the river bed topography has a continuous gradient between adjacent sample points (Sear and Milne, 2000). Then the grid corners with estimated elevations in the curvilinear coordinate are converted to the initial Cartesian coordinate system which is illustrated in Figure 2.4. After transforming sample locations in Cartesian coordinates into the curvilinear coordinate, the river topography can be estimated by commonly used spatial interpolation methods. Some interpolation methods were selected to estimate river topographies in the curvilinear coordinate, including linear interpolation, cubic spline interpolation, natural neighbour interpolation, nearest-neighbour interpolation and IDW. The results are presented in the Appendix.

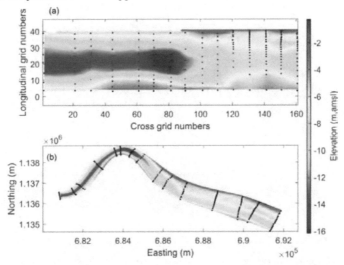

Figure 2.4. The estimated bathymetric elevations by the triangle-based interpolation (a) and these elevations converted to a Cartesian coordination system (b).

2.2.3 Calibration of ordinary kriging

To carry out an OK interpolation, nugget, sill, range and model type parameters are empirically defined. This may lead to unexpected errors. To deal with this problem, we combined the Monte Carlo approach with OK interpolation to optimize selecting uncertain factors. As mentioned in the OK section, the semivariogram function qualifies the spatial correlation of depth samples. From the spatial correlation analysis, the parameters are selected by a fitting empirical semivariogram. Therefore, the Monte Carlo approach is used. This approach is to randomly select a value of selected parameters from the semivariogram outputs. The selected parameters are sill, nugget and range and model type. Figure 2.5 depicts the semivariogram model of the interpolated samples. As a result, a fitting Spherical model is empirically defined in which the nugget, partial sill and range are 5, 10 and 8000, respectively. By applying the Monte Carlo approach, the kriging interpolation is implemented in a large number of 1000 iterations. The partial sill, nugget and range are selected in ranges of the model's fitting values adding/subtracting a half of these values. After each iteration, the correlation coefficient was recorded for each interpolated dataset.

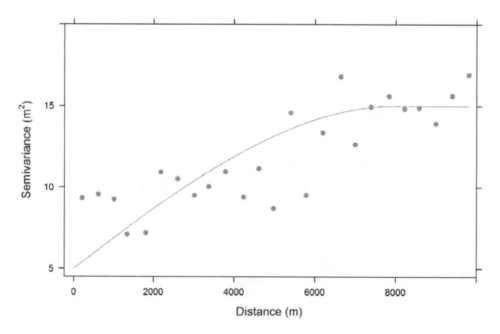

Figure 2.5. The cross-section data (points) and the fitting Spherical model (line). The nugget, partial sill and range are 5, 10 and 8000, respectively.

2.2.4 Performance assessment

Cross-validation is commonly conducted to validate spatial interpolation methods (Curtarelli et al., 2015; Zhang et al., 2016) . In general, the cross-validation is efficient in the cases that the number of validating samples is much lower than the number of training samples. However, in the situation that the number of validating samples is really larger than the number of training samples, cross-validation is unnecessary.

The performance of the used interpolation methods is assessed by calculating errors which can be presented by statistical indices. There are numerous indicators for determining performance of interpolation methods. The three indices of coefficient of correlation (R), root mean square error (RMSE), and bias are chosen to assess interpolation method performance. R is the degree of relationship between estimation and observation. An R value of 1 is the perfect correlation and indicates that estimation and observation are equal. RMSE is an accuracy measure (Walther and Moore, 2005) which is the standard deviation of the interpolation errors. A bias measure used is mean error (ME) which is computed by mean difference between observations and estimates. The R, RMSE and ME are computed as below.

$$R = \frac{n(\sum s.o) - (\sum s)(\sum o)}{\sqrt{[n\sum s^2 - (\sum s)^2][n\sum o^2 - (\sum o)^2]}} \quad (3)$$

$$RMSE = \sqrt{\frac{\sum_1^n (s-o)^2}{n}} \quad (4)$$

$$ME = \bar{s} - \bar{o} \quad (5)$$

where s is estimation, o is measurement and n is a number of samples.

2.3 RESULTS AND DISCUSSION

2.3.1 Calibration of ordinary kriging

The results of sensitivity analysis help to understand relationship of semivariogram parameters and estimation accuracy and choose the optimal values of these parameters. Figure 2.6 presents the results of sensitivity analysis in which correlations of OK predictions and observations are interpreted by the nugget, sill, range and model type parameters over 1000 iterations. The OK interpolation method is a stable interpolator which estimates unsampled values in a reasonable agreement with measured data. R values vary in a range from 0.5 to 0.75. R and nugget inversely relate, but R and partial sill are positively related. Among the parameter sets in sensitivity analysis, the optimal values of nugget and partial sill are 2 and 13 m2. This sill value of 15 is equal to the values of the empirical fitting curve. It is found that the sill is easily defined when the spatial correlation of data starts levelling. Nonetheless, the optimal nugget in 1000 iterations is 2, smaller than that of the empirical fitting curve. Thus when analysing the data, the nugget parameter is more difficult to define than the sill. As a result of the analysis, the mentioned empirical leads to the best R of 0.7 while the optimal R increasing to 0.75. In the sensitivity analysis, the random distances are selected from 4,000 m to 12,000 m. In fact, when the distances increase, the semivariances increase as well. This means that when pairs of sample points are at large distance, they have less correlation. Thus the range has a negative relationship with R. The optimal range is 4,250 m. This distance is similar to the length of around 4 cross-section data. It means that river bed topography has relation with the 4 closest cross-section data. In the cases that the samples are uniformly or regularly distributed in space, OK is one of the best interpolators (e.g Huang et al. 2015). However, spatial distribution of the cross-section data is clumped and river bed topography is influenced by its flow directions. Thus the OK in Cartesian coordinates is not capable of predicting river bed surfaces based on limited cross-sectional data. Among the selected model types, interpolators with the spherical model result in the highest R, peaking at 0.75. This indicates that the spherical function is the best fitting model in this analysing data. The spherical models are commonly chosen for interpolating river bed topography (e.g. Carter and Shankar, 1997; Zhang et al., 2016).

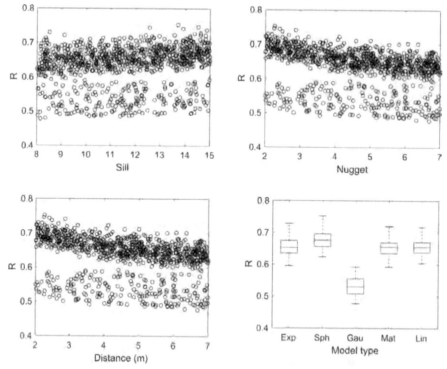

Figure 2.6. Sensitivity analysis of semivariogram's parameters.

2.3.2 Interpolation of river bed topography

Figure 2.7 depicts interpolating results of AR, OK, KED and IDW interpolating methods compared to the measured data. The river bed topography generated by the OK method is the worst surface in comparison to measured data. It cannot produce the cross-section shape. For instance, the western segment is a relatively flat bed and the thalweg is not clearly visualized as in the measured data. Therefore, it is not capable of capturing the general trend of river bed topography. In contrast, the three other methods generated topographies in which the thalweg is reasonably captured. However, prediction errors are different among these methods. The IDW interpolated topography has a slightly discontinuous thalweg and a number of jags. These jags obviously appear in the middle of cross-sections where the values of river bed topography are equally affected by the two cross-sections. In order to reduce errors of the discontinuous thalweg problem, the distance-to-thalweg factor was taken into account of the Kriging with External Drift interpolator (KED). This approach is efficient in generating the thalweg in estimated topography. Nevertheless, this topography has several artificial dunes along the river. These dunes are apparent at shallow cross-sections. This discrepancy may be resulted from the geostatistical approach. For river bed topography, it is better to apply interpolation methods that assume a continuous gradient between sample points of cross-sections. This is determined by the AR interpolated topography. After reducing anisotropy effects, the topography is predicted by the linear interpolation method. The results of AR interpolation approach have a good agreement with measured data as it is able to capture the continuous thalweg and predict river bed topography accurately. Nonetheless, there are differences in elevation especially at the southern river bank. The river bank elevation is difficult to reproduce because the values of samples used in interpolating dataset are not high as of measured samples. In an application of hydrodynamic modeling, these errors of river bank elevation unlikely influence hydrodynamic results because the river banks are usually dry areas. The difficulties of bank elevations can be overcome by adding data from digital terrain models and considering into interpolation.

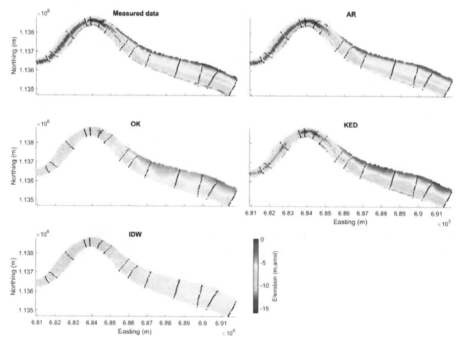

Figure 2.7. River bed surface estimations of the Cua Tieu branch by different interpolation methods of anisotropy-removed interpolation (AR), Ordinary Kriging (OK), Kriging with External Drift (KED) and Inverse Distance Weighting (IDW). The black dots are samples which are used for interpolation.

2.3.3 Interpolation method validation

To quantify the accuracies of the applied method, the discrepancies between predictions and measurements are interpreted by the indices of R, RMSE and ME. Figure 2.8 shows scatterplots of predicted and measured topographies and the values of validating indices. The AR method has the highest R of 0.97, followed by the IDW, KED and OK. These R values indicate that the relationship between estimated and measured values have a strong correlation. This determines that the AR method has better performance to simulate variations of river bed topography. However, R cannot present the difference between predictions and measurements. Thus, ME is a common bias measure computed to quantify a system error of these methods. Generally, these methods are good estimators, except for the OK. The AR, IDW and KED generate unbiased results, with absolute ME values smaller than 0.1 m while the ME value of the OK is about 0.37 m. Then RMSE is used to define accuracy of these predictors. An accurate method should be precise as it predicts unsampled points with small variations. The highest RMSE values of approximately 2 m are for the OK and IDW estimator. The KED have values of 1.69 m. The smallest variation is generated by the AR method, with RMSE of 0.74 m. By these validating indices, it concluded that the AR method is a good estimator in predicting river bed topography from cross-section data.

Although the AR is a good estimator, it still has a certain error. To give an insight into further studies, spatial distribution of errors is depicted in Figure 2.9. It clearly show that a high frequency of error samples occurs in areas where the river sides and bottom are linked (around 100-200 m from the river banks). This holds for all four interpolating methods. As a reason, the river in this case has U-shaped cross-sections so elevations in this area are highly variable. Therefore, this characteristic should be noted when defining representative cross-sections for interpolation.

The AR approach includes a combination of three steps and we found that the main step for efficiently generating the river channel topography is coordinate conversion in which samples in Cartesian coordinates are transformed into curvilinear coordinates based on the river banks and the thalweg. For instance, the topography of the Tieu branch in the curvilinear coordinate was estimated by some common interpolation methods. The results, are shown in Figure 2.12 in the Appendix, have a good agreement with measured data. These interpolation methods in the curvilinear coordinate generate accurate topographic results. Noticeably, the IDW interpolator in the curvilinear coordinate performs a better topography compared to in Catesian coordinates.

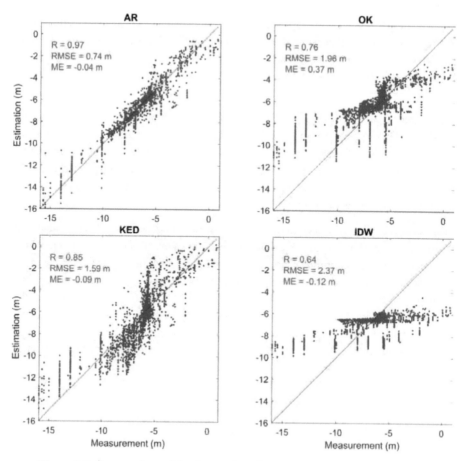

Figure 2.8. Scatter plots of depth samples of measurements and predictions by interpolation methods of anisotropy-removed interpolation (AR), Ordinary Kriging (OK), Kriging with External Drift (KED) and Inverse Distance Weighting (IDW). Additionally, the performance of these methods are interpreted by indices of coefficient of correlation (R), root mean square error (RMSE) and mean error (ME).

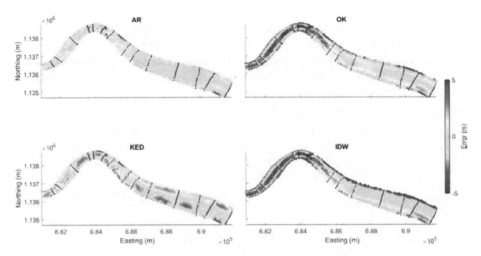

Figure 2.9. Spatial distribution of errors generated by interpolators of anisotropy-removed interpolation (AR), Ordinary Kriging (OK), Kriging with External Drift (KED) and Inverse Distance Weighting (IDW).

2.3.4 Comparison of cross-sections and thalweg line

Figure 2.10 presents measurement and interpolation of the three selected cross-sections. These cross-section were selected based on river widths, representing as narrow, medium and wide cross-sections. These cross-sections are at the middle of the two adjacent cross-sections used for interpolation, because the middle cross-sections have less influences of these two adjacent cross-sections. In general, the cross-sections reconstructed by using OK and IWD are flat and unrealistic. It obviously shows effects of OK and IWD approaches. Specifically, this appeared to be caused by isotropic approaches for interpolating river topography. In fact, when the directional characteristic of the river topography was considered in KED interpolation, the cross-section shapes were reproduced. The KED could be able to reproduce the shapes of cross-sections, but it could generate cross-section elevation precisely (Figure 2.10). The AR approach is capable of reconstructing cross-sections shape and elevation accurately. For instance, various morphological singularities of the selected cross-sections were reproduced by using the AR approach. However, there are slight discrepancies between measured and reconstructed elevations. These discrepancies could be appeared when the samples used for interpolation do not include morphological variations. This should be noted for field measurement of river topography.

Figure 2.11 shows the thalweg profiles of measured data and topography reconstructed by AR, IDW, OK and KED approaches. The IDW and OK dramatically underestimate thalweg elevations, but they can produce the slope of bed surface. The KED considering the thalweg line also resulted in an unsatisfactory thalweg elevations. Although the KED reasonably generated the thalweg line at the deepest region, it overestimated the fluctuation of the thalweg. This led to non-realistic dunes and ripples in the interpolated bed surface. The AR approach generated a good fit thalweg. It is able to capture the variations of the river bed surface. Reconstructing thalweg line is highly sensitive to cross-sectional inputs. For example, the deepest part is unlikely reproduced if its adjacent cross-sections are used as inputs.

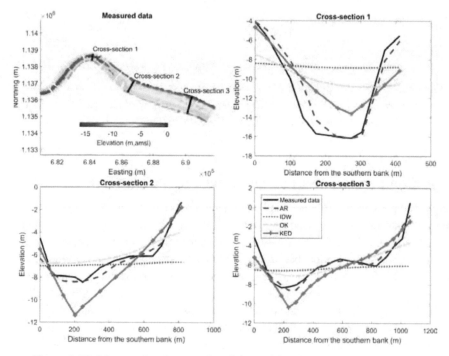

Figure 2.10. Measured and interpolated data of the three selected cross-sections.

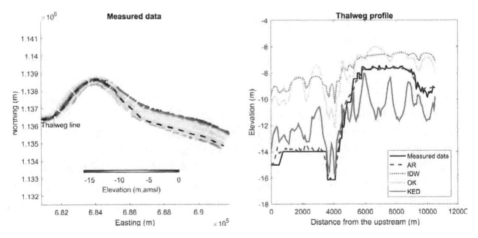

Figure 2.11. Measured and interpolated data of the thalweg line.

Channel thalweg, as the natural direction of a watercourse, is an important factor to reconstruct river topography. Incorporating the thalweg into the interpolator results in a continuous deepest channel. This makes the bed surface reconstructed more accurate. We introduced an efficient interpolation approach for generating river bathymetry from sparse cross-sectional data. We also found that the linear interpolation method is suggested for sparse data regions. However, it is difficult to identify the thalweg line based on sparse data. We suggest to generate the thalweg line by connecting splines of the deepest points from cross-section to cross-section. Besides, there is several studies considering the center line for reconstructing river bathymetry (Goff and Nordfjord, 2004; Legleiter and Kyriakidis, 2008; Merwade, 2009). This could lead to generating a discontinuous thalweg channel. Chen and Liu (2017) used the interpolation methods, namely linear interpolation, IDW and Natual Neighbor to resample cross-sections and found that the linear interpolation is the most efficient method to reproduce smooth topography and continuous thalweg trajectory.

2.4 CONCLUSIONS

We proposed an efficient and accurate interpolation approach to be applied for cases of sparse data of river topography. The performance of this method is tested in the sparse data and validated with a high spatial resolution of the Tieu estuarine branch. This approach has better performance in comparison with the commonly used interpolation methods of IDW, OK and KED. The major difference between these methods is that a channel-fitted coordinate was incorporated, so this is an essential step to reconstruct river bed topography accurately.

From the results of this study, two major conclusions for interpolating river topographies are drawn. First, excluding anisotropic effects of river topographies should be implemented before applying spatial interpolation methods. River banks and thalwegs are necessary references to diminish the anisotropic effects, especially in meandering rivers. Second, linear interpolation method is one of the best methods to produce river topography from cross-sections. This method generates smooth bed surfaces which are better for hydrodynamic modeling. This study found that reducing anisotropic effects of river channel topography is the main step to reproduce river topography and suggests to convert from Cartesian coordinates to the curvilinear coordinate.

The AR approach for generating river bed topography is helpful in cases of data-poor regions. It is more important when river bed topography plays a driven role in numerical modeling. Moreover, this approach is crucial to generate topography of missing data cases. The AR is quite flexible compared to some commercial GIS software. Further work will consist of extension in order to generate banks elevations from LiDAR data or laser scanning.

APPENDIX

In this section the results of some selected interpolation methods were obtained in the transformed coordinate. The theory of these interpolation methods are described by de Boor (2001), Mitas and Mitasova (2005) and Webster and Oliver (2007).

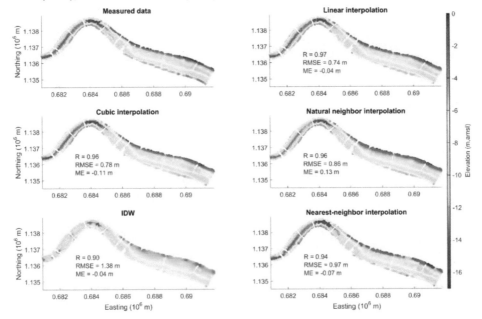

Figure 2.12. Topographies of the Tieu branch interpolated by selected methods in the transformed coordinate in comparison with measured data. The interpolated topographies generated by each method are evaluated by statistical indices of correlation (R), root mean squared error (RMSE) and mean error (ME).

3

MULTI-SCALE SEDIMENT MODELING[2]

Abstract

Fluvial sediment is the major source for the formation and development of the Mekong Delta. This chapter aims to analyse the dynamics of suspended sediment and to investigate the roles of different processes in order to explore flux pattern changes. We applied modeling on two scales, comprising a large-scale model (the whole delta) to consider the upstream characteristics, particularly the Tonle Sap Lake's flood regulation, and a smaller-scale model (tidal rivers and shelf) to understand the sediment processes on the subaqueous delta. A comprehensive comparison to in-situ measurements and remote sensing data demonstrated that the model is capable of qualitatively simulating sediment dynamics on the subaqueous delta. It estimates that the Mekong River supplied an amount of 41.5 mil tons from April 2014 to April 2015. A substantial amount of sediment delivered by the Mekong River is deposited in front of the river mouths in the high flow season and resuspended in the low flow season. A sensitivity analysis shows that waves, baroclinic effects and bed composition strongly influence suspended sediment distribution and transport on the shelf. Waves in particular play an essential role in sediment resuspension. The development of this model is an important step towards an operational model for scientific and engineering applications, since the model is capable of predicting tidal propagation and discharge distribution through the main branches, and in predicting the seasonal SSC and erosion/deposition patterns on the shelf, while it is forced by readily available inputs: discharge at Kratie (Cambodia), GFS winds, ERA40 reanalysis waves, and TPXO 8v1 HR tidal forcing.

---

[2] This chapter is based on:

Thanh, V. Q., Reyns, J., Wackerman, C., Eidam, E. F. and Roelvink, D.: Modelling suspended sediment dynamics on the subaqueous delta of the Mekong River, Cont. Shelf Res., 147(August), 213–230, doi:10.1016/j.csr.2017.07.013, 2017.

3.1 INTRODUCTION

Deltas are low-lying plains with both riverine and marine influences (Nguyen Anh Duc, 2008). In these coastal environments, saline and fresh water, and riverine and marine sediments mix. As such, deltas are nutrient-rich, and very productive ecological areas with high biodiversity. Deltas have socio-economic importance, as they support up to ~500 million people (Syvitski et al., 2009). Consequently, deltas are heavily impacted by human activities such as land-use changes, port development, land reclamation, diking, damming of channels, dredging and sand mining. These anthropogenic factors add to natural fluctuations and lead to modifications of hydrodynamics (Marineau and Wright, 2014), sediment discharge (Manh et al., 2015; Renaud et al., 2013), and morphology (Dissanayake and Wurpts, 2013). Rivers have been used as a fresh water resource for agricultural irrigation and hydropower dams, leading to changes in river morphology and hydrodynamics (Anthony et al., 2015). In addition, deltas are highly vulnerable to climate change, including sea level rise and other natural hazards (Renaud et al., 2013; Wong et al., 2014). According to recent assessments, 40 deltas in the world are projected to be at risk in terms of coastal erosion due to a decrease of sediment supply and due to sea level rise in combination with subsidence (Wong et al., 2014). Among these, the Mekong Delta is one of the deltas that are extremely vulnerable to predicted sea level rise by the year 2050 (Tri et al., 2013). The main reason for this vulnerability is that deltas are highly dynamic areas that can be significantly impacted by changing environmental conditions. They are low-lying areas (about 26,000 km2 in the world lying below local mean sea level), which makes them vulnerable to sea level rise (Overeem and Syvitski, 2009). Understanding systematic sediment dynamics of a delta is vital to assess impacts of the mentioned factors.

In this study we focus on the Vietnamese Mekong Delta (VMD, Figure *3.1*). It is home to 17 million people and has been extremely modified by human interventions, such as hydropower dams and sand mining (Anthony et al., 2015; Brunier et al., 2014; Cosslett and Cosslett, 2014). Different approaches have been applied to understand sediment dynamics in the VMD, including field data analyses (e.g. Hung et al., 2014a; Nowacki et al., 2015; Wolanski et al., 1996), processing of satellite images (e.g. Heege et al., 2014; Loisel et al., 2014) and numerical modeling (e.g. Xue et al., 2012; Manh et al., 2015; Vinh et al., 2016). Among these approaches, modeling is a powerful tool for understanding sediment dynamics in a complex system like the VMD, while the others could be used for validating modeling work. Furthermore, modeling is not only helpful in explaining sediment dynamics, but also in predicting bed composition changes under changing forcing and boundary conditions.

A number of modeling studies have been carried out in order to understand hydrodynamics and sediment transport in the Mekong Delta. For modeling on the scale of the whole delta, Wassmann et al. (2004) conducted a study to assess impacts of sea level rise on rice production by using a 1D numerical model (VRSAP). Results show that inundation in the VMD would shift significantly landward due to the projected sea level rise. It can be concluded that the tide plays a significant role in the flood drainage in the VMD. A similar inundated pattern was found by Van et al. (2012). They also applied a 1D hydrodynamic model (ISIS) which was used by the MRC (Mekong River Commission) to study changes of flood characteristics under upstream development and sea level rise. As a result, flooding caused by tides would be stronger due to sea level rise impacts, and may lead to changes in sediment convergence near the river mouths. Manh et al. (2014) estimated the spatial sediment distribution in the Mekong Delta by using a MIKE11 model to simulate rivers (1D) and floodplains (quasi-2D). The study indicated that more than half (53%) of the sediment is transported to the VMD's coastal areas via the Can Tho and My Thuan distributaries, but the sediment dynamics in these areas have not been investigated yet. Xing et al. (2017) investigated sand dynamics of the Song Hau branch and morphological changes. Their model suggests that the ratio between seaward and landward velocity will increase and morphology would be highly affected in the context of relative sea level rise and delta plain subsidence. For the marine realm, Xue et al. (2012) modelled the sediment budget for the Mekong Delta shelf by applying the COAWST system (Warner et al., 2010). It was found that ~90% of total sediment derived from the Mekong River is concentrated in the delta front zone. Although their study considered interactions of wave, tides, and currents, the dynamical role of river mouths that are the connections of river and ocean were not considered. Regarding the modeling of connections of river and ocean, Vinh et al. (2016) modelled the variation of suspended sediment concentration (SSC) as a function of the time-varying wave climate by using a numerical model (Delft3D). Their domain covered the Mekong River from Can Tho (on the Song Hau) and My Thuan (on the Song Tien) to the sea. Noticeably, the increase of wave height would enhance sediment resuspension and expand the sediment plume seaward along the delta front. The study is attractive in that it systematically explores the sediment concentration and transport patterns as a function of the river discharge and wind/wave conditions. However, since their study was applied for high flow and low flow seasons separately, it could not reflect an important characteristic of sediment dynamics on the shelf: during the high-flow season, a large amount of sediment coming from the Mekong River is transported and deposited on the adjacent shelf (near river mouths), and this sediment is resuspended and transported south-westward due to seasonal currents and waves (Hein et al., 2013).

Therefore, the aims of this study are to analyse the dynamics of suspended sediment and to investigate the roles of different processes leading to flux pattern changes using physics-based models on hourly to seasonal timescales, and spatial scales ranging from the entire VMD to a mangrove fringe. The models, which are forced by easily obtainable open source data, are the key component toward developing an operational model train

31

that will allow scientists and engineers to investigate the effects of changing forcing and/or engineering works on sediment dynamics.

3.2 METHODOLOGY

For optimized modeling of a large area such as the Mekong Delta, multiscale modeling has been used on two levels so far. The large scale Delft3D Flexible Mesh (Deltares, 2016) model is used to generate boundary conditions for a coastal area Delft3D-4 model, and its domain includes all main Mekong branches from Kratie, Cambodia to the Vietnamese East (South China) Sea (Figure 3.1). The model is extended upriver to Kratie because Kratie is the first main gauging station outside the tide-influenced zone. Based on measured water levels at Kratie, the annual floods start when water levels increase over 20.9 m (MRC, 2007). The water regime in the delta is strongly regulated by the Tonle Sap Lake levels. In this regard, water discharge from the Mekong River upstream of Kratie is stored in the lake, which acts as a natural flood regulator for the Mekong Delta, and includes the southern parts of Cambodia and the VMD (DHI, 2004). Additionally, the lake can store approximately half of the total inflow from the Mekong River and then release ~90% of outflow to the Mekong River (RoyalHaskoningDHV et al., 2010). Therefore, the Tonle Sap Lake regulation is considered in the modeling.

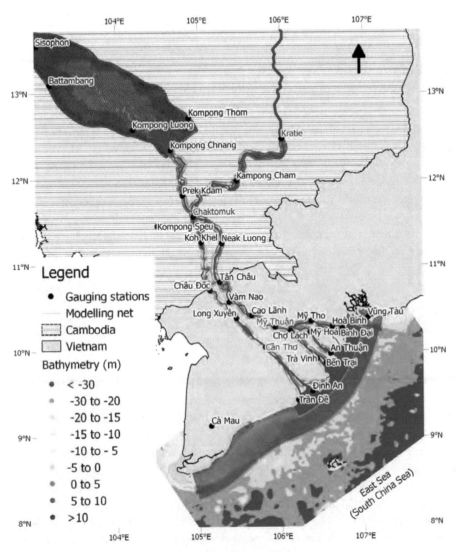

Figure 3.1. River bathymetry from cross-section interpolation and shelf bathymetry from GEBCO (Weatherall et al., 2015) of the Mekong Delta, model network, and measurement stations.

The small scale (Delft3D-4) model (Lesser et al., 2004) has a finer grid resolution and focuses only on the estuaries of the main branches, the coastal areas, and the shelf. This model allows investigations of the mechanisms of sediment dispersal on the subaqueous delta due to coastal processes on a seasonal timescale. Hence, the small scale model has

33

a fine grid resolution to better resolve the local flow characteristics, especially in the estuaries. In order to optimize the computational time, this model is reduced in spatial extent, covering the Song Hau and the Song Tien from Chau Doc and My Thuan respectively, to the Vietnamese East Sea. The model extent on the shelf is approximately 70 km, which is enough to contain the sediment plumes completely.

The multiscale approach used in this study has advantages in large domains, such as the Mekong Delta. Two versions of the physical-based models, namely Delft3D FM and Delft3D-4 were used simultaneously to simulate hydrodynamics of the Mekong River while suspended sediment transport is only computed by the Delft3D-4 model. The Delft3D-4 model is capable of simulating the hydrodynamics and suspended sediment transport for the whole Mekong Delta, but Delft3D-4 only calculates on structured grids. Thus, applying the Delft3D FM model, with mixed structured and unstructured grids, would be a preferred solution. However, the sediment module of Delft3D FM is not validated yet. Therefore, we used these two models to optimize the computations.

3.3 MODEL DESCRIPTION

3.3.1 Delft3D-4

Delft3D-4 (structured, curvilinear version 4, as opposed to the unstructured, flexible mesh version Delft3D-FM) is an integrated flow and transport modeling system that has been developed by Deltares (Lesser et al., 2004). It can simulate two dimensional (depth-averaged) or three dimensional unsteady flow and transport phenomena (including sediment, temperature, salinity and passive tracers). It contains modules of flow, water quality, ecology, waves and morphology. The Delft3D flow module solves the unsteady shallow water equations (Deltares, 2014). In the horizontal, a curvilinear approach is applied. In the vertical, the user can choose from grid systems that use sigma coordinates or fixed Z coordinates. The wave module used in Delft3D is the third generation SWAN model (Booij et al., 1999). The wave module can be coupled with the flow module in two ways (online and offline coupling). Recently, Delft3D-4 has been widely used in different applications in hydrodynamics (e.g. Garcia et al., 2015; Hasan et al., 2012; Lu et al., 2015), sediment transport (Gebrehiwot et al., 2014; Hu et al., 2009; Vinh et al., 2016), and morphology (e.g. Dastgheib et al., 2008; Duong et al., 2012; van der Wegen et al., 2011; Van der Wegen and Roelvink, 2012; Xing et al., 2017).

3.3.2 Delft3D Flexible Mesh (FM)

In this study an unstructured hydrodynamic model, Delft3D FM, has been applied to simulate hydrodynamics and suspended sediment transport in the large scale. It is a 1D-2D-3D model developed by Deltares, calculating on unstructured meshes, with different shape of meshes, including triangles, rectangles, and pentagons (Deltares, 2020b). It solves the 2-D and 3-D shallow water equations using finite volume techniques. Deltares (2016); Kernkamp et al., (2011) and Martyr-Koller et al., (2017) describe the conceptual model in detail. Generally, Delft3D FM and Delft3D-4 are similar; the most noticeable difference between Delft3D-FLOW and Delft3D FM is the use of structured and unstructured grids respectively. Although Delft3D FM still is under continuous development, several studies have been implemented using Delft3D FM in salinity and offline-coupling suspended sediment modeling (e.g. Achete et al., 2015).

3.4 MODEL SETUP

The land boundaries that are used for defining the modeling domains have been digitized from a freely accessible satellite imagery (Google, 2016). There is some bathymetry data available for the coastal and ocean areas of the Vietnamese East Sea such as GEBCO (Weatherall et al., 2015) and ETOPO (Amante and Eakins, 2009). However, the resolutions of these datasets (approximately 1 x 1 km2) are not sufficient for the numerical modeling of rivers. For this reason, these data are used only for the open sea parts of the domain. The river bathymetry has been generated by interpolating the underlying cross-sectional data of a 1D model (Van et al., 2012). Additionally, the bathymetry of coastal areas, particularly in the river mouths, was significantly updated by recent measured data (Vo Luong, 2016, pers.comm.).

3.4.1 Set up of Delft3D FM

In the large scale model, unstructured grids containing orthogonal quadrangle and triangle shapes are used to schematize the complex geometry of the Mekong River. Following suggestions of Kernkamp et al. (2011), curvilinear-like geometry has been used for rivers and sea areas while river junctions, river mouths, and complicated meandering rivers are schematized by triangular nets for convenient connectivity. This strategy for grid creation enhances computational accuracy. The model contains highly variable river widths and a complex coastal zone, so the grid resolution is also highly spatially variable; average resolutions range between 100-200 m, 300-600 m and 1000 m for connecting rivers, main river branches, and sea areas, respectively. The large scale model used the 2D depth-averaged setting. The upstream boundary at Kratie is defined by the total water discharge, fed by data collected from the telemetry data of the local gauging station and the weekly situation reports published by Mekong River Commission (MRC, 2016). The downstream boundary conditions are composed of astronomic components, extracted as the 8 principal tidal constituents from the global tide model TPXO 8v1 (Egbert and Erofeeva, 2002). On the lateral offshore boundaries, a Neumann boundary condition is applied in order to allow free development of cross-shore water level slopes and flow profiles due to effects of tides, Coriolis, wind, and waves (Roelvink and Walstra, 2004). These prescribed alongshore water level gradients are determined from the seaward alongshore water level boundary conditions. This model excludes effects of waves since it is applied to simulate river discharge distribution. Water levels in the model fluctuate considerably and the Tonle Sap Lake plays a significant role in controlling the upstream discharge in the low flow season. Therefore, the correct specification of the initial conditions is important in order to reduce the required spin-up time of the simulations (Ji, 2008). The bottom roughness is prescribed by spatial distributions of Manning coefficients, ranging from 0.032 to 0.016. The range of roughness coefficients is adapted from Manh et al. (2014) and their distribution resembles the ones used by Vinh et al. (2016).

3.4.2 Set up of Delft3D-4

The computational domain and bathymetry used in the small scale model are presented in Figure 3.2. The hydrodynamics and sediment transport of the small scale model is modelled using a 3-D simulation with 10 vertical boundary-fitted sigma layers. The boundary conditions were defined as total discharges on river boundaries, water levels along the shore parallel offshore boundary and Neumann boundaries for the lateral offshore boundaries (Figure 3.3). The river boundaries are obtained by nesting into the large scale model. Tidal forcing is similar to the large-scale model; wave data on the offshore boundaries are derived from ERA Interim reanalysis data (http://apps.ecmwf.int/datasets/data/interim-full-daily) extracted at point (lat,lon)=(9°,106.5°).

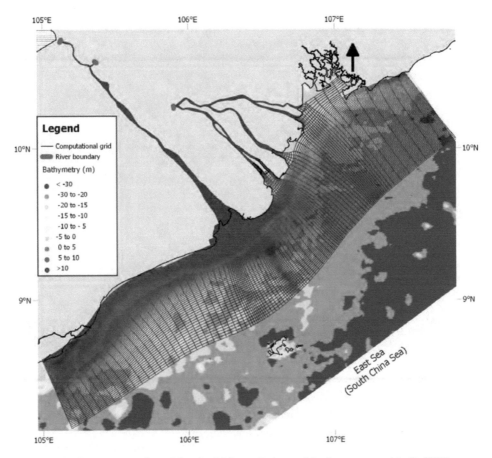

Figure 3.2. Computational grid for the Mekong Delta and bathymetry used in Delft3D-4.

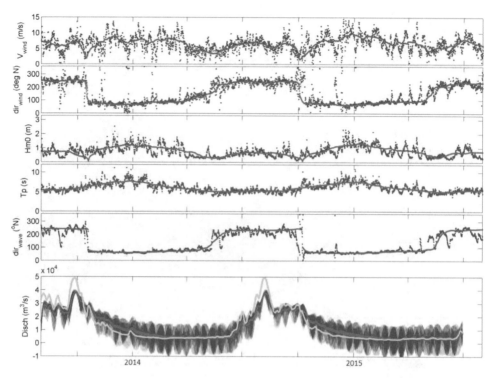

Figure 3.3. Time series of wind speed and direction, wave height (Hm0), wave period (Tp) and wave direction and total discharge used for forcing the Delft3D-4 model. In the bottom panel the green line indicates the discharge at Kratie; the blue line the sum of discharges computed by Delft3D-FM and imposed on the Delft3D-4 model. Red dots in top five panels are running averages over a one-month period; in the bottom panel over a 7-day period. The difference in discharge between Kratie and the sum of the Mekong branches is due to buffering in Tonle Sap.

Suspended sediment in the Mekong Delta is fine sand/silt or cohesive, flocculating material. According to measurements reported by Koehnken (2014), 98% of the total suspended sediment load (by volume) in Tan Chau (for location, see Figure 3.1) is silt (2-63 μm) and clay (<2 μm). The suspended sediment characteristics vary spatially in the delta. Grain sizes are generally coarser upstream, and finer downstream (Gugliotta et al., 2017). The median grain sizes d_{50} are 15 μm and 3.9 μm in the upstream (near Tan Chau) and downstream estuary reaches, respectively (Hung et al., 2014b; Wolanski et al., 1996). Nevertheless, the sediment grain size may locally be larger due to flocculation. The flocculated grain size of sediment was 40 μm based on analysis of Wolanski et al. (1996). Recently, a study of sediment grain size in the Song Hau (Stephens et al., 2017) showed that median grain sizes range from 5 μm to 20 μm. Therefore, the uniform sediment grain size d_{50} of 15 μm is used in our modeling. Additional sediment modeling parameters for the fresh water areas are derived from extensive field-based experiments conducted by Hung et al. (2014b). For saline reaches, sediment parameters are derived from Vinh et al. (2016). Specifically, critical bed shear stresses for deposition τ_{cbd} and erosion τ_{cbe} are 1000 Pa (implying continuous deposition) and 0.2 Pa, respectively. This range covers the bed shear stresses, which were generally <6 Pa and <2 Pa near the Song Hau mouth during ebb and flood tides, respectively, in the high flow season (Mclachlan et al., 2017). The constant erosion rate M is 2 x 10-5 kg/m²/s. Furthermore, settling velocities of sediment in fresh and salt water are 0.05 and 0.325 mm/s, respectively. According to observations of Mclachlan et al. (2017), the specific particle settling velocities ranged from 0.01 to 1 mm/s, varying substantially between ebb and flood tides, and in high flow and low flow seasons.

3.5 MODEL VALIDATION

3.5.1 Model calibration

The large scale model (Delft3D FM)

The large scale model calibration was performed for the year of 2000 as a comprehensive data set was available in order to compare the simulated characteristics in detail. The relative mean absolute error (RMAE) is selected as an error statistic, following Sutherland et al. (2004). The RMAE was calculated based on Equation 6.

$$RMAE = \frac{(|S-M|)}{|M|} \quad (6)$$

where S and M are the mean of simulated and measured water levels at a station respectively.

Water levels in the VMD stations sharply change between high flow and low flow seasons so monthly RMAE values of 17 stations were calculated with the purpose of classifying error variations due to effects of the seasonal changes. Typically, two-thirds of RMAE values are in excellent and good classifications (Table 3.2 and Table 3.3 in Appendix) according to Sutherland et al. (2004)'s scheme, indicating good agreement of simulations and observations. Nonetheless, results for some stations (Can Tho, My Tho, and My Thuan) in flood months are in the poor and bad categories. This result, most likely due to the effect of incomplete bathymetry data, indicates a need for further model improvement.

Simulated and measured water levels and discharges in 2000 are illustrated in Figure 3.4 and Figure 3.5. Generally, the water levels reveal good agreement between simulation and observation in terms of amplitude and phase. Although there are still absolute errors at the troughs of the water level time series, the model correctly simulates tidal propagation in the main branches of Mekong River. Throughout the year 2000, the simulated discharges are overestimated in the high flow season, in particular, at the Vam Nao station (in the flood zone) due to the lack of floodplain and lateral river system schematizations. However, the model has better skill in the low flow season when the water volumes are spatially constrained to the river bed. This result suggests that the floodplain and lateral river systems should be included for better model performance.

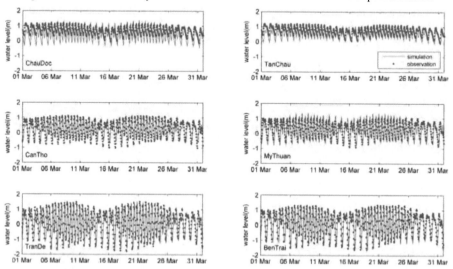

Figure 3.4. Model performance of water levels at selected stations in the VMD in 2000. Left and right panels are along the Song Hau and Song Tien respectively. The top to bottom panels present comparisons at stations in the upstream to river mouths of the VMD.

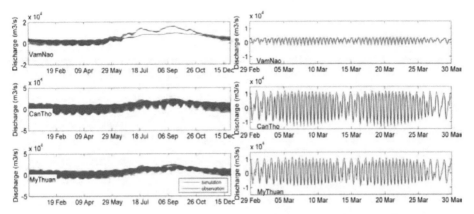

Figure 3.5. Comparison of simulated and measured water discharges (m³/s) at Vam Nao, Can Tho, and My Thuan stations over the year 2000 (left panels) and in March 2000 (right panels).

The small scale model (Delft3D-4)

Since the objective of the present modeling exercise is to understand the mechanisms of seasonal sediment dynamics, modelled periods selected included the high flow and low flow seasons. The small scale model was used to model the high flow season of 2014 and the low flow season of 2015. The RMAE values of Chau Doc, Vam Nao, Can Tho, My Thuan and Vam Kenh stations are 1, 0.58, 0.02, 0.53 and 0.24, respectively. These indices reveal that the model performance is generally reasonable. The results are also evaluated by the Skill value (Equation 7):

$$Skill = 1 - \frac{\sum_{i=1}^{N}(|S_i - \bar{M}_i|)^2}{\sum_{i=1}^{N}(|S_i - \bar{M}_i| - |M_i - \bar{M}_i|)^2} \qquad (7)$$

where S and M are simulated and measured water levels, respectively, i is the hourly time step, and N is the number of simulated hours. The Skill values of the 5 stations range from 0.85 to 0.95. In fact, the model closely reproduces the water levels as presented in Figure 3.6.

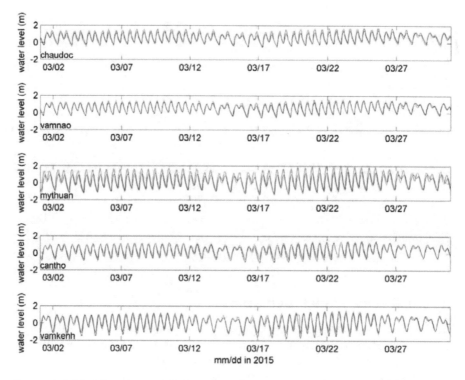

Figure 3.6. Model performance of water levels at the main stations in the VMD in 2015. Red and blue lines are simulated and measured water levels, respectively.

To assess the general behaviour of the 3D distribution of the salinity over one-year simulation, we analysed the monthly maximum salinity in the bottom layer (Figure 3.7) and the monthly minimum salinity in the top layer (Figure 3.8). The first gives an indication of the seasonal variability of the saline intrusion into the branches and shows maximum saline intrusion in March extending just beyond the bifurcation in the Song Hau; this is somewhat short of the observed intrusion length reported by Nowacki et al. (2015) and Mclachlan et al. (2017), likely due to incomplete representation of the deeper channels. However, the seasonal variation agrees well with observations (Figure 3.9).

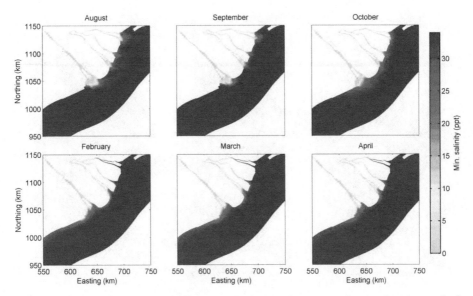

Figure 3.7. Monthly maximum salinity in the bottom layer over the simulated period of 2014-2015.

The extent of the freshwater plume is illustrated by the minimum salinity in the top layer per month, and is obviously highly correlated with the discharge. Also, the plumes from all branches are of similar importance and interact with each other, a clear reason to include all major branches in the modeling. The plume is well contained within the model and the flexible boundary conditions at the lateral NE and SW boundaries allow the south-west going plume during the low flow season to exit the model undisturbed. The seasonal pattern as expected shows north-eastward flow in August and September; the plume is more variable in October and clearly shifts to south-westward flow in the low flow season.

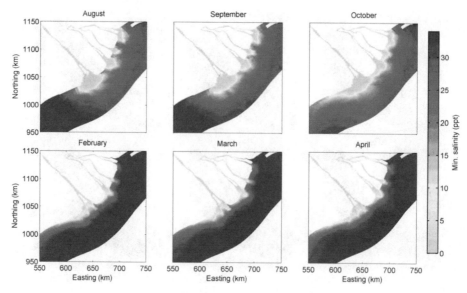

Figure 3.8. Minimum salinity in the top layer over the simulated period of 2014-2015.

For suspended sediment calibration, simulated results were temporally and spatially compared to observed data. To calibrate SSC temporally, the simulated results were compared to data collected by offshore in-situ measurements (Eidam et al., 2017). Vertical structure of suspended sediment data was collected by an optical backscatter sensor (OBS) during the high flow season of 2014 (29 Sept 2014) in front of the mouth of the Song Hau (Figure 3.9). The simulated SSC and data from OBS is presented in Figure 3.9, bottom panels. Absolute values of the sediment concentration differ for various reasons, so at this stage we focus on the qualitative comparison between the vertical distributions of observed and modelled sediment concentration throughout the tidal cycle. There is a good agreement in the time of bottom sediment resuspension, and the height to which this sediment is resuspended. At this depth and observation period, the effect of wave stirring is not very important, as is shown by the results for the 'no waves' run. The surface water is clearly identified in Figure 3.9, top panels, with a correct representation of the less saline surface plume, which contains a significant concentration of sediment, both in observations and in the model. Although the resolution of 10 equidistant sigma layers and a background vertical viscosity of $10\text{-}4$ m^2/s may lead to some diffusion, the patterns are still relatively sharp both in salinity and sediment, and qualitatively correct. Obviously, as shown for the results without salinity, this process is not represented without considering the 3D stratification.

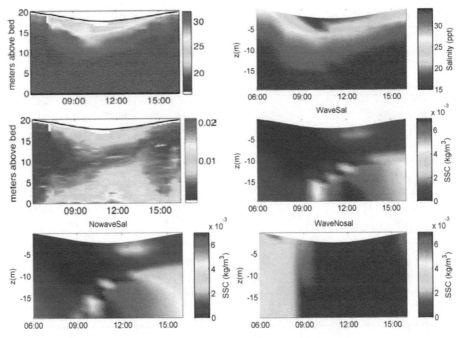

Figure 3.9. Top panels: vertical structure of measured (left) and simulated (right) salinity in front of the Song Hau mouth on 29ᵗʰ Sept 2014. Middle panels: vertical structure of OBS response (left) and simulated sediment concentration (right) in front of the Song Hau mouth on 29ᵗʰ Sept 2014. Bottom panels: same simulations without waves (left) and without effect of salinity (right).

3.5.2 Model validation

In order to validate the spatial variability of the modelled SSC fields, we compared the numerical results with selected satellite images collected by Wackerman et al. (2017) in which there were distinct features related to the sediment plume from the Mekong branches, both for the high flow season and the low flow season. Table 3.1 gives an overview of the images and the hydrodynamic conditions during each image. The patterns of simulated river plumes were compared to remotely sensed images in Figure 3.10 (high flow) and Figure 3.11 (low flow). The model is generally able to model the river plumes in the both high flow and low flow seasons. The simulated patterns of river plume extents and alongshore SSC are compatible with observations of satellite images. Specifically, the river plumes tend to the north-east in the high flow season Figure 3.10. Moreover, the alongshore SSC patterns consistently appeared in both modelled results and satellite images. It is noticed that though the patterns are similar, the scales are not the same in satellite-derived SSC and model results. Justification for this may be uncertainty in the penetration depth for the remote sensing observations.

Table 3.1. Overview of satellite images and conditions during their capture.

	Sensor	Year	Month	Day	hour	min	Water level (m)	Discharge (m³/s)	Wave height (m)
1	Landsat 8	2014	9	2	3	14	0.1	17875	0.5
2	Aqua MYD02HKM	2014	9	23	6	0	1	736	0.7
3	Terra MOD02HKM	2014	9	24	3	40	-0.4	22851	0.5
4	Terra	2015	3	3	3	40	-0.4	9704	0.9
5	Terra MOD02HKM	2015	3	5	3	30	-1.1	17398	0.6
6	Rapid Eye	2015	3	17	4	10	0.6	-12517	0.6

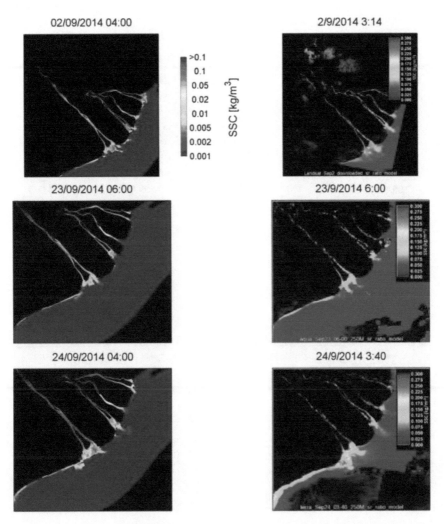

Figure 3.10. Simulated SSC (left panels) and satellite images (right panels) collected and converted to SSC values (kg/m³) using the ratio method (source: Wackerman et al., 2017) during high flow season.

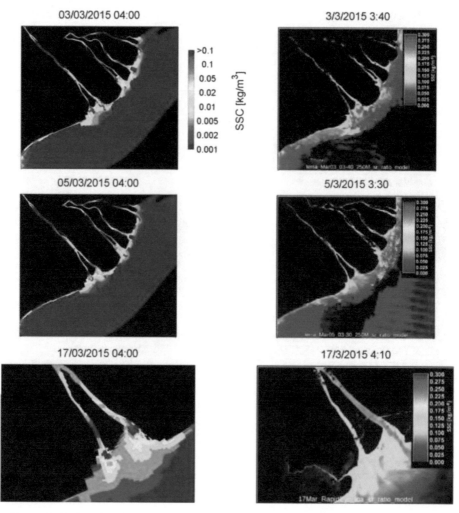

Figure 3.11. Simulated SSC (left panels) and satellite images (right panels) collected and converted to SSC values (kg/m³) using the ratio method (source: Wackerman et al., 2017) during low flow season.

3.6 RESULTS AND DISCUSSION

3.6.1 Seasonal sediment dynamics

In this study, the seasonal sediment dynamics are presented based on river flows, tides and waves. The former are classified as low flow and high flow seasons to compare variations of riverine sediment export to the sea. The tides are based on tide variations, encompassing spring and neap tides as maximum and minimum tidal ranges; for each tidal variation, suspended sediment in the surface layer is shown at high water slack (HWS) and the following low water slack (LWS) when sediments can have a small and large seaward extent, respectively. The waves are strong in the winter monsoon (the high flow season), with wave heights of about 1m. Waves are smaller in the summer monsoon (the low flow season) during which heights decrease to 0.5 m.

For the high flow season, distributions of sediment concentration are illustrated in Figure 3.12. The farthest offshore sediment extent of the Song Hau was found at LWS time of spring tide in the high flow season in which the contours of 0.05 kg/m^3 sediment concentration can reach as far as 20 km from the Song Hau mouth. Its dispersal is to north-eastward directions due to the constraint of coastal currents. The flux patterns reflect the interactions of river flow and coastal currents (Unverricht et al., 2013). In contrast, small amounts of suspended sediment are exported beyond the river mouths when water level is the highest during neap tides of the high flow season. The SSC dynamics during tidal cycles were simultaneously found by McLachlan et al. (2017), with the highest SSC (> 0.2 kg/m^3) near the river mouth during spring tides and relatively smaller SSC of < 0.2 kg/m^3 during neap tides.

For the low flow season, Figure 3.13 presents sediment concentrations during spring tide and neap tide. Sediment concentrations in the river and shelf are low, except in the nearshore region, because SSCs from the upstream (near Vam Nao station) are ~50% lower than in the high flow season (Hung et al., 2014a). A considerable source of alongshore sediment transport in the low flow season comes from the bed sediment layer that was formed in the high flow season (Szczuciński et al., 2013; Xue et al., 2010). This processed result is demonstrated in Figure 3.14.

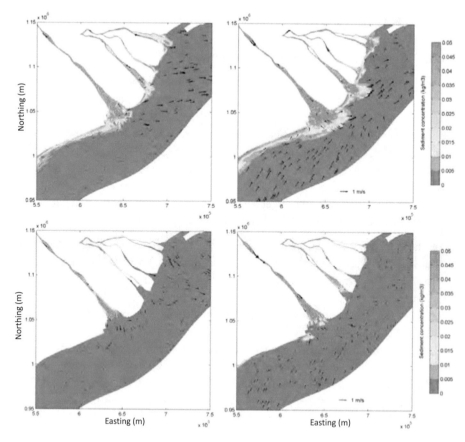

Figure 3.12. Simulated sediment concentration in the surface layer at HWS (left) and LWS (right) of spring tide (top panels) and neap tide (bottom panels) in the high flow season. The arrows are surface currents.

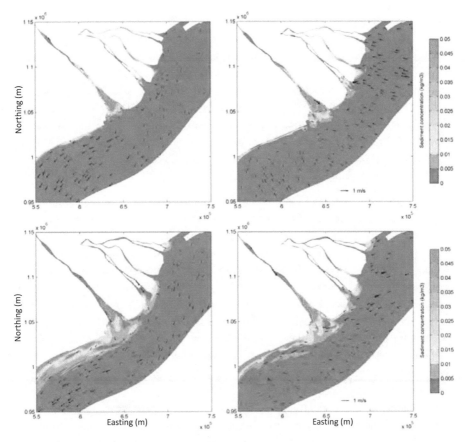

Figure 3.13. Simulated sediment concentration in the surface layer at HWS (left) and LWS (right) of spring tide (top panels) and neap tide (bottom panels) in the low flow season. The arrows are surface currents.

3.6.2 Sediment flux

The totally cumulative sediment load through the river boundaries from April 2014 to April 2015 is estimated to be ~41.5 million tons. High flow season (June to October) sediment loads account for 90% and low flow season (November to May) sediment loads account for only 10% of the total. The sediment load is approximately a quarter of the commonly used estimate of 160 million tons/year (Milliman and Syvitski, 1992). However, recent estimates were slightly lower, i.e., ~40 million tons/year in the Mekong mouths (Nowacki et al., 2015). Noticeably, sediment load into the delta (through Tan Chau and Chau Doc) was only over a half of total sediment load at Kratie (Manh et al.,

2014). According to recent measurements by Lu et al. (2014), sediment discharge flowing into the delta at Phnom Penh (in the Mekong mainstream, near the Tonle Sap confluence) averaged 67 million tons/year from 2008 to 2010. The amount of sediment is trapped by floodplains and river systems, so the remaining part, around 53% of the total sediment load at Kratie, is transported through Can Tho and My Thuan (Manh et al., 2014). Manh et al. (2014) presented that the sediment discharge through Can Tho and My Thuan ranges from 25.9 to 50.5 million tons/year in low- and high-flow years. In this paper, an estimated 31 million tons/year are accumulatively transported via Can Tho and My Thuan. Furthermore, the ratio of sediment export of the Song Hau and Song Tien in the high flow season closely fits with the rates calculated by Manh et al. (2014).

Figure 3.14 presents spatial distribution of cumulative sediment transport through the cross-sections and sedimentation in the high flow and low flow seasons. The spatial sediment transport has a similar pattern compared to Xue et al. (2012) and Vinh et al. (2015) in general. Here, we describe the seasonal sediment transport processes on the subaqueous delta based on the model. In the high flow season, sediment transport is dominated by the river due to high water discharge from the river, when 90% of annual sediment volume is discharged. Alongshore suspended sediment on the subaqueous delta is transported north-eastward from the Song Hau mouth to the Song Tien mouth, accounting for 4% of the total riverine sediment supply, while the remaining fluvial sediment load is deposited in front of the river mouths (Figure 3.14) due to onshore currents induced in bottom layers (Xue et al., 2012). In the low flow season, much less sediment is exported from the river, but the alongshore sediment transport rate is nine times higher than the sum of riverine sediment outputs. The strongest bottom stress occurs due to waves and seasonal currents in this period, leading to resuspension of the sediment layer that was deposited in the last flood season. Wave heights in this season are high, fluctuating around 1 m and twice higher than in the high flow season (Figure 3.3; Eidam et al., 2017). This process results in strong alongshore sediment transport south-westward.

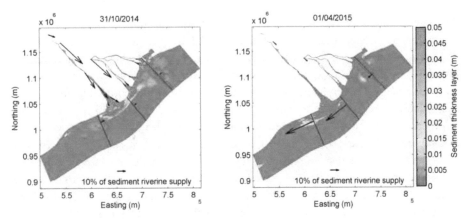

Figure 3.14. Sediment thickness layers at the end of the high flow season (left) and low flow season (right). The arrows indicate cumulative suspended sediment transport through cross-sections over the high flow and low flow season.

3.6.3 Sensitivity to salinity, waves and sediment processes

In this section we investigate the sensitivity of the seasonal behaviour of the sediment on the shelf in response to salinity gradients, waves, and the bed sediment composition, through analysis of monthly averaged patterns. This allows for a qualitative comparison with the satellite data analysis (Figure 3.15) presented by Loisel et al. (2014). Additionally, we look at time series of the sediment volume integrated over a number of control areas. We carried out 4 simulations spanning a year (1 Apr 2014 thru 1 Apr 2015), which included both measurement campaigns and the most important part of the seasonal cycle. The forcing conditions were as shown in Figure 3.3. Four simulations were carried out:

1. WavesSal: a base simulation with one mud fraction with a uniform fall velocity of 0.0325 m/s and no sand; with wave forcing and 3D salinity effects.

2. NoWavesSal: same as WavesSal but with wave forcing turned off

3. WavesNoSal: same as WavesSal but with uniform salinity

4. WavesSalSand: same as WavesSal but with a more complex representation of sediment processes: an initial 2 m thick sand layer with d50 of 0.11 mm and (on top of that) an initial mud layer of 1 mm with fall velocity ranging from 0.05 mm/s in fresh water to 0.325 mm/s in saline water, cf. Vinh et al. (2016). The bed composition in this case was described by a transport layer of 10 cm thickness, 20 layers of 10 cm thickness, and a flexible underlayer.

53

The base simulation WavesSal (Figure 3.16) shows patterns rather similar to the multi-year monthly averaged results shown in Loisel et al. (2014). In general the region of high SSCs remains near to the coast. September and October show the highest concentrations in the estuary mouths as a result of the high-discharge, sediment-laden flood flows. As the wind pattern shifts in September towards north-easterly directions, the plume turns southward and hugs the coast. The sediment deposited during the high flow season, mostly on the edge of the delta front, is strongly resuspended by waves and currents in December and January. The strongly reduced riverine output in February, March, and April and the depletion of the sedimentation areas leads to strongly reduced concentrations throughout, though nearshore SSC remains relatively high due to the combined effect of wave stirring and the cross-shore gravitational circulation with an onshore-directed near-bed current. A striking resemblance with the observed patterns is found in the increase of suspended sediment towards the south of the Ca Mau Cape.

The sensitivity of the SSC patterns to the different scenarios is illustrated for the months of September (the high flow season), January and March (the low flow season, Figure 3.17). The effect of waves is clearly visible in the comparison between WavesSal and NoWavesSal; concentrations in the nearshore area and on the delta front are generally lower without the additional stirring effect of the waves.

The effect of no salinity gradients is mainly to spread the sediment farther over the shelf, due to two mechanisms: the sediment is spread more evenly throughout the vertical water column since there is no reduction in the vertical diffusivity due to buoyancy effects, and there is no near-bed onshore gravitational circulation. Together these effects lead to a marked reduction in the nearshore SSC.

In the WaveSalSand simulation, where the sandy bed layers are added, the longshore extent of the sediment plume is sharply decreased. SSC is substantially concentrated in front of and in between of the river mouths. In addition, the nearshore SSC is negligible even when the strongest currents occur in December and January. This is caused by the effect of the initially mostly sandy mixing layer, where resuspension of the mud is proportional to the fraction of mud present; it takes time for an equilibrium bed composition to develop, which happens quickly on the delta front but probably takes several years further away from the mouths. Therefore, although the description of sediment for this run is more complete, the SSC patterns on a larger scale are less reliable than the WavesSal simulation with a single fraction.

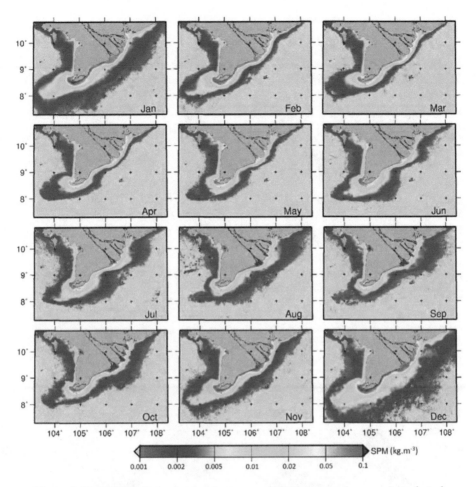

Figure 3.15. MERIS-derived monthly averaged SSC patterns (source: Loisel et al., 2014).

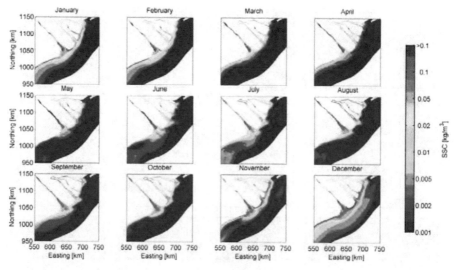

Figure 3.16. Spatial distribution of monthly mean SSC in the WavesSal simulation.

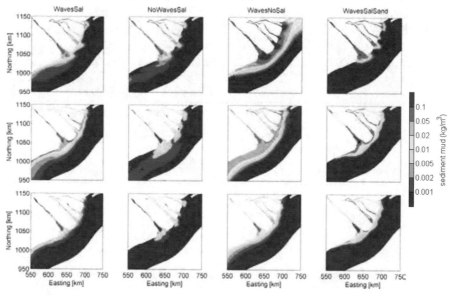

Figure 3.17. Sensitivity of spatial distribution of monthly mean SSC in September 2014 (top panels), in January 2015 (middle panels) and in March 2015 (bottom panels).

3.6.4 Temporal-spatial distribution of at-bed sediment volume

To further illustrate the sedimentation/erosion behaviour on hourly to seasonal timescales we analyse the time series of at-bed fine sediment volume integrated over the areas indicated in Figure 3.18. These areas were classified, referring to the regional burial rates calculation in DeMaster et al. (2017). These areas includes the areas of the Song Hau mouth (BassacMouth), the mouth of the main Song Tien's branch (MekongMouth) and the mouth of the other Song Tien's branches (SaigonMouth). The shelf was divided into 3 areas of the northern proximal (ShelfNorth), the southern proximal (ShelfMid) and the central-distal (ShelfSouth).

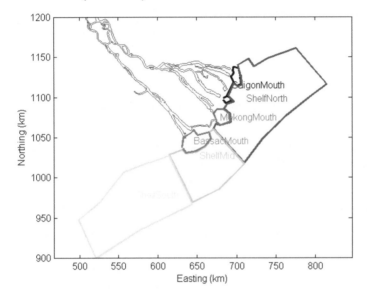

Figure 3.18. Control areas for sediment volume calculations.

The resulting time series of total volume per control area are shown in Figure 3.19. In the base case, WavesSal, the sediment volumes in the river mouths increase during the high flow season, in a relatively smooth way; during October they remain more or less constant and then, as wave action increases and discharges decrease, the deposited material is resuspended and transported southward. In the simulations with only a mud fraction, all material including the 1 mm initial mud layer is eroded away and net sedimentation is negative, or, if there had not been any sediment to begin with, would have been zero. There is definitely some fine sediment in the bed, so this does not appear realistic, and is one reason why the simulation with an additional sand layer was executed. For that simulation, the mud deposited in the mouths is mixed with sand in the mixing layer and thus becomes more difficult to erode; hence between one-third and two-thirds of the deposited mud remains in the mouths, mostly at the edge of the delta front. In the shelf control areas, most of the sedimentation takes place in the mid shelf area in front of the mouths; also here all sediment is eroded away. In the north shelf the balance is around zero during the high flow season and then all is eroded during the low flow season and northeast monsoon. The south shelf area fluctuates around zero during the time that sediment passes by from the north and due to local resuspension and deposition during and after wave events, respectively. With the sand layer added, these areas turn into sinks for the fine sediment and all that is deposited remains there. Clearly reality must be somewhere between the WavesSal and WavesSalSand simulations.

Turning off wave effects has a notable impact in the river mouths, where the erosion after the high flow season is much less pronounced and a small net sedimentation remains. On the shelf the effect is less pronounced but still significant: the north shelf erodes less than with waves, and the remaining sediment does not make it to the south shelf.

The effect of salinity gradients is to keep the sediment much closer to the coast. The sedimentation in the mouths is less and therefore there is more sedimentation on the shelf when the salinity gradients disappeared.

We may conclude from these results that though the general trends of seasonal deposition and erosion are realistic, the role of the bed composition processes on the concentration patterns and on the net sedimentation or erosion is very significant. For further studies, we need to consider these in more detail and through much longer simulations.

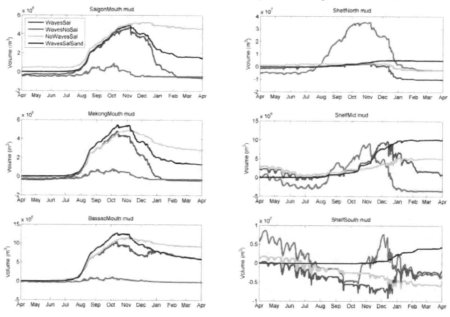

Figure 3.19. Sediment volume variation of control areas.

3.7 CONCLUSIONS

We applied process-based models to simulate present-day sediment transport processes in the Mekong delta. The multiscale modeling approach combines accurate modeling of the seasonal and interannual variation of the water and sediment fluxes throughout the entire Mekong delta with more detailed 3D modeling of the estuarine and shelf processes. Although there are some discrepancies between simulation and observation, this may be improved by adding lateral channel networks and the floodplain system. With regard to sediment and salinity dynamics, the model simulates the vertical distribution of salinity reasonably well, compared to in-situ data. Moreover, the detailed model snapshots of the sediment plumes show reasonable agreement with satellite-derived SSC fields from Wackerman et al. (2017), and monthly averaged SSC patterns are in general agreement with satellite-derived monthly climatologies presented by Loisel et al. (2014).

Seasonal sediment transport changes are strongly modulated by river discharges and monsoons. The seasonal sediment dynamics can be summarised as follows: (i) in the high flow season, the Mekong River delivers a large amount of sediment that is deposited on the delta front due to coastal processes (waves and currents); (ii) in the low flow season, while delivered volumes of riverine sediment substantially decrease, waves and currents have an important role in resuspending the sediment layer formed during the former flood season. Furthermore, the sediment movement depends on the direction of the monsoon, which determines the sediment plume dispersal patterns.

Three additional simulations were carried out to analyse effects of salinity, waves and sediment composition on sediment dynamics. The sensitivity analysis revealed that without wave effects, the nearshore SSC is considerably reduced in absence of wave-induced stirring. The SSC field expands farther seaward in conditions of constant salinity gradients because there is no decrease of vertical diffusivity due to buoyancy effects and no gravitational circulation. In the simulation wherein sand fractions are added, the suspended sediment concentrations primarily fluctuate in the river mouth fronts and the alongshore sediment transport is reduced. This fluctuation is a numerical morphological spin-up effect as the bed composition needs time to develop into a well-mixed bed, a period during which excessive amounts of mud are trapped within the mixing layer. This result highlights the need to carry out multi-year simulations, using morphological acceleration and input reduction techniques where needed.

APPENDIX

Table 3.2. Relative Mean Absolute Error values of water levels along the Song Hau

Station	ChauDoc	VamNao	LongXuyen	CanTho	DaiNgai	TranDe	DinhAn
Jan	0.511	0.422	0.422	0.194	0.182	0.204	0.431
Feb	0.493	0.425	0.412	0.155	0.174	0.171	0.395
Mar	0.404	0.328	0.293	0.011	0.110	0.052	0.296
Apr	0.173	0.032	0.116	0.383	0.198	0.165	0.087
May	0.018	0.216	0.353	0.649	0.337	0.279	0.016
Jun	0.153	0.404	0.571	1.051	0.528	0.415	0.143
Jul	0.247	0.556	0.760	1.680	0.711	0.417	0.127
Aug	0.061	0.247	0.431	1.625	0.736	0.505	0.228
Sep	0.057	0.338	0.483	1.492	0.559	0.228	0.037
Oct	0.177	0.033	0.080	0.695	0.179	0.065	0.184
Nov	0.392	0.238	0.247	0.005	0.235	0.233	0.448
Dec	0.401	0.256	0.253	0.042	0.183	0.174	

Error classification of Sutherland et al. (2004)

< 0.2   Excellent

0.2-0.4   Good

0.4-0.7   Reasonable

0.7-1.0   Poor

>1   Bad

Table 3.3. Relative Mean Absolute Error values of water levels along the Song Tien

Station	Tan Chau	My Thuan	My Hoa	My Tho	Tra Vinh	Hoa Binh	Ben Trai	An Thuan	Binh Dai	Vam Kenh
Jan	0.412	0.411	0.263	0.193	0.260	0.118	0.271	0.271	0.188	0.036
Feb	0.382	0.379	0.231	0.167	0.208	0.071	0.208	0.247	0.156	0.011
Mar	0.260	0.229	0.138	0.102	0.114	0.018	0.144	0.158	0.076	0.023
Apr	0.051	0.201	0.206	0.231	0.204	0.249	0.171	0.135	0.221	0.306
May	0.206	0.492	0.392	0.452	0.359	0.443	0.305	0.296	0.356	0.467
Jun	0.258	0.798	0.597	0.786	0.503	0.644	0.417	0.456	0.545	0.664
Jul	0.332	1.150	0.725	1.029	0.533	0.781	0.422	0.453	0.564	0.708
Aug	0.098	0.936	0.743	1.172	0.537	0.821	0.433	0.460	0.600	0.735
Sep	0.213	0.622	0.461	0.885	0.182	0.494	0.133	0.160	0.301	0.421
Oct	0.007	0.114	0.127	0.365	0.076	0.211	0.053	0.011	0.083	0.244
Nov	0.214	0.233	0.276	0.086	0.375	0.133	0.367	0.349	0.241	0.103
Dec	0.239	0.269	0.213	0.086	0.271	0.086	0.299	0.265	0.199	0.037

4

FLOODING IN THE MEKONG DELTA: THE IMPACT OF DYKE SYSTEMS ON DOWNSTREAM HYDRODYNAMICS[3]

Abstract

Building high dykes is a common measure of coping with floods and plays an important role in agricultural management in the Vietnamese Mekong Delta. However, the construction of high dykes causes considerable changes in hydrodynamics of the Mekong River. This paper aims to assess the impact of the high-dyke system on water level fluctuations and tidal propagation in the Mekong River branches. We developed a coupled 1-D to 2-D unstructured grid using Delft3D Flexible Mesh software. The model domain covered the Mekong Delta extending to the East (South China Sea) and West (Gulf of Thailand) seas, while the scenarios included the presence of high dykes in the Long Xuyen Quadrangle (LXQ), the Plain of Reeds (PoR) and the Trans-Bassac regions. The model was calibrated for the year 2000 high-flow season. Results show that the inclusion of high dykes changes the percentages of seaward outflow through the different Mekong branches and slightly redistributes flow over the low-flow and high-flow seasons. The LXQ and PoR high dykes result in an increase in the daily mean water levels and a decrease in the tidal amplitudes in their adjacent river branches. Moreover, the different high-dyke systems not only have an influence on the hydrodynamics in their own branch,

[3] This chapter is based on:

Thanh, V. Q., Roelvink, D., van Der Wegen, M., Reyns, J., Kernkamp, H., Vinh, G. Van and Linh, V. T. P.: Flooding in the Mekong Delta: the impact of dyke systems on downstream hydrodynamics, Hydrol. Earth Syst. Sci., 24, 189–212, 2020.

but also influence other branches due to the Vam Nao connecting channel. These conclusions also hold for the extreme flood scenarios of 1981 and 1991 that had larger peak flows but smaller flood volumes. Peak flood water levels in the Mekong Delta in 1981 and 1991 are comparable to the 2000 flood as peak floods decrease and elongate due to upstream flooding in Cambodia. Future studies will focus on sediment pathways and distribution as well as climate change impact assessment.

4.1 INTRODUCTION

Rivers are the major source of fresh water for human use (Syvitski and Kettner, 2011). In addition, the fresh water supply is an important resource for ecosystems. When river discharge exceeds the bankfull discharge, the floodplains inundate. Fluvial floods cause both advantages and disadvantages for local residents. Floods are the main source of fresh water supply and deliver sediments that act as a natural and valuable fertiliser source for agricultural crops (Chapman and Darby, 2016). This is an important process in the Mekong Delta as the majority of local citizens are farmers. In contrast, extreme floods may damage both crops and infrastructure.

In order to maintain agricultural cultivation during the high-flow seasons, dyke rings have been built to protect agricultural crops in the Vietnamese Mekong Delta (VMD). As a result, the river system in the VMD has significantly changed, especially following the severe floods in 2000 (Biggs et al., 2009; Renaud and Kuenzer, 2012). A dense canal system has been created in flood-prone areas to efficiently drain flood waters from the Long Xuyen Quadrangle and the Plain of Reeds to the West Sea (Gulf of Thailand) and to the Vamco River respectively (Figure 4.1).

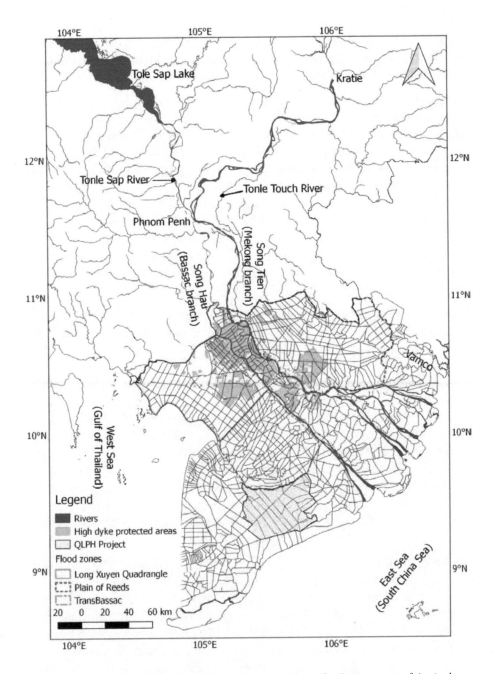

Figure 4.1. Location of the Mekong Delta. Adapted from the Department of Agriculture and Rural Development of An Giang and Dong Thap (2012).

65

Recently, large hydraulic structures have been built not only in the flood-prone areas but also in coastal areas to protect cropping systems from saline intrusion. Therefore, hydrodynamic processes have changed considerably. Understanding the prevailing hydrodynamics is essential for sustainable water management in these regions.

The high-dyke system is intended to reduce local natural flood hazards but may alter the hazard downstream (Triet et al., 2017). Furthermore, this system also increases the potential risk due to dyke breaks. Following different approaches, Tran et al. (2018) found that the high-dyke system upstream of the VMD causes an increase in the peak water levels in downstream areas. However, water levels at these downstream stations are dominated by tidal motion. In fact, tides may result in an increase in water levels in the central VMD. Thus, an analysis of tidal fluctuation is needed to investigate water level changes on the Mekong River. The high-dyke system may be an important factor, but sea level rise in combination with land subsidence enhances peak water levels at the central stations to a larger extent (Triet et al., 2017). The high-dyke system influences not only the downstream hydrodynamics by reducing inundated floodplain areas, but also impacts fluvial sediment deposition on floodplains.

There are a number of large-scale numerical models used to simulate the annual floods and suspended sediment transport and to evaluate the impacts of dyke construction in the Mekong Delta (Manh et al., 2014; Tran et al., 2018; Triet et al., 2017; Van et al., 2012; Wassmann et al., 2004). For instance, Tran et al. (2018) investigated the impacts of the upstream high-dyke system on the downstream part of the VMD. Using a MIKE hydrodynamic model for the Mekong Delta, they found that the high-dyke system in the Long Xuyen Quadrangle (LXQ) can reduce the discharge of the Tien River, diverting around 7 % of the total volume to the Hau River. In addition, the yearly discharge variations have slight effects on the peak water levels at the Can Tho station, while Triet et al. (2017) found that the high-dyke system caused an increase from 9 to 13 cm in the flood peaks at the central VMD stations. Moreover, Triet et al. (2017) showed that the development of the dyke system upstream of the VMD reduced flood retention in this area, leading to an increase of 13.5 and 8.1 cm in the peak water levels in the downstream part of the VMD at Can Tho and My Thuan respectively.

The above-mentioned studies evaluated the impact of high dykes in the LXQ and the high dykes developed up until 2011. However, the impacts of the other floodplain regions need to be considered, including the LXQ, the Plain of Reeds (PoR) and Trans-Bassac. Additionally, Manh et al. (2014), Tran et al. (2018), and Triet et al. (2017) used a common 1-D version of the MIKE11 model for the Mekong Delta, and the downstream boundaries are defined at the Mekong's river mouths. However, Kuang et al. (2017) found that river flows can contribute to a rise in the water level at the river mouths. Thus, in the present study another modeling approach is used in order to address these issues.

This study aims to assess the impacts of the high-dyke system on water level fluctuation and tidal propagation in the Mekong River branches. An unstructured, combined 1-D to 2-D grid is used to simulate the flood dynamics in 2000. The model domain covers the Mekong Delta and extends from Kratie in Cambodia to the East (South China Sea) and West (Gulf of Thailand) seas. Simulated scenarios present the impact of high dykes in different floodplain regions and the entire VMD. The specific objectives are

- to develop a calibrated and validated hydrodynamic model using Delft3D Flexible Mesh that is able to simulate the annual floods in the Mekong Delta;

- to analyse the spatial–temporal distribution of the Mekong River's flows for different extreme river flow scenarios; and

- to evaluate how the development of high dykes, which are built to protect floodplains, influences the downstream hydrodynamics, particularly with respect to tidal propagation.

4.1.1 The Mekong Delta

The Mekong is one the largest rivers in the world (MRC, 2010). It starts in Tibet (China) and flows through five riparian countries before reaching the ocean via (originally nine branches but now) seven estuaries. It has a length of 4800 km and a total catchment area of 795 000 km^2 (MRC, 2005). The Mekong Delta starts in Phnom Penh (Figure 4.1), where the Mekong River is separated into two branches, namely the Mekong and the Bassac (Gupta and Liew, 2007; Renaud et al., 2013). The Mekong Delta is formed by sediment deposition from the Mekong River, which provides an annual water volume of 416 km^3 as well as 73 Mt yr^{-1} of sediment at Kratie, which is mainly distributed in the high-flow season (Koehnken, 2014; MRC, 2005). The Mekong Delta has a complex river network, especially in the Vietnamese region. The Mekong Delta's river network is illustrated in Figure 4.2. It has resulted from extensive artificial canal development that began in 1819 (Hung, 2011).

Regarding land resources, the VMD area comprises about 4×10^6 ha, and three-quarters of this region is used for agricultural production (Kakonen, 2008). The livelihoods of the local citizens are primarily based on agriculture and aquaculture. Thus, the river infrastructure has been developing with agriculture as a priority. The area provides just over half of the rice yields in Vietnam and provides up to approximately 90 % of the exported rice yields from Vietnam (GSOVN, 2010). However, the rice cultivation is highly influenced by annual floods (MRC, 2009a).

The most intensive agricultural production in the VMD is found in An Giang Province (Figure 4.2). Although it is also a flood-prone area, the inundation periods are slightly shorter due to flood withdraw to the West Sea. In the deep flooded zones (the Long Xuyen Quadrangle and the Plain of Reeds), high dykes have been densely constructed in the downstream direction. This is due to the fact that the areas downstream of the LXQ and PoR experience low flood peaks, meaning that the dyke rings do not need to be as high as they do in the regions upstream of the LXQ and PoR.

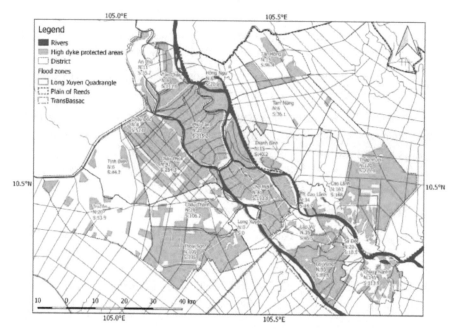

Figure 4.2. Spatial distribution of high dykes that were built up until 2011. The high dykes are presented by district. The number of high dykes (N) is shown as well as the area that they protect (S) in square kilometres (km²). Adapted from the Department of Agriculture and Rural Development of An Giang and Dong Thap (2012).

The Mekong Delta is dominated by a tropical monsoonal climate. There are two dominant monsoons. The southwest monsoon occurs from May to October, coinciding with the high-flow season. The other, drier monsoon period is from November to March and is followed by a transition period (MRC, 2010). The mean temperature is approximately 26.5 °C. Although the climate is seasonally changing, monthly average temperature differences are 4 °C between the hottest and coldest months (Lê Sâm, 1996). However, seasonal rainfall is drastically different in terms of time and space. The high-flow season contributes approximately 90 % of the total annual rainfall intensity, whereas the low-flow season (from December to April) accounts for 10 % of the total rainfall. The yearly mean rainfall is about 1,600 mm in the VMD. The highest rainfall is found in the western coastal area of the Mekong Delta and ranges between 2,000 and 2,400 mm. The eastern coast receives about 1,600 mm of rainfall, whereas the lowest rainfall is recorded in the centre of the VMD (Lê Sâm, 1996; Thanh et al., 2014).

4.1.2 High-dyke development in the Vietnamese Mekong Delta

The Mekong Delta has been modified extensively over the last 2 decades following the devastating flood in 2000. One noticeable change is the hydraulic infrastructure, especially the dyke development. Before the dykes were built, a dense canal network was developed to drain floods to the West Sea and to clean acid sulfate soils.

Depending on the dyke function, dykes can be classified into two categories. Low dykes are built to protect the rice harvest from summer–autumn crops in August. This is the rising phase of the annual floods. The low dykes allow flood overflows and the inundation of floodplains, so the crests of low dykes are designed to just equal the maximum water level in August. High dykes are constructed in order to completely prevent the annual floods and enable intensive agricultural production. Generally high dykes are designed at a crest level of 0.5 m above the year 2000 flood peak. The 2000 flood was a severe event that has a 50-year recurrence interval in terms of flood volume (MRC, 2005). In An Giang, there are two types of high dykes. The first type of high dyke only has a single dyke ring. The hydrodynamics just outside of the dyke ring are dominated by floods. These dykes have a straightforward floodplain protection function but a high risk of breaching. The other type of high dyke contains the above-mentioned single dyke ring which is then protected by a large outer dyke. The hydrodynamics outside of these high dykes are controlled by structures (sluice gates) in the outer dyke.

Several studies have mapped the high dykes in the VMD using remotely sensed images (e.g. Duong et al., 2016; Fujihara et al., 2016; Kuenzer et al., 2013). Using this method, the high dykes are identified via flooded and non-flooded areas. However, these results are easily affected by the water management of the high-dyke rings. For example, in An Giang, the high-dyke areas are managed according to the 3+3+2 cropping cycle rule. In other words, these areas are cultivated for eight (3+3+2) agricultural crops over 3 years and are allowed to inundate during part of the year once every 3 years. Thus, the results need to be verified with observations to ensure the reliability of the maps.

High dykes were rarely constructed in the VMD before 2000 (Duong et al., 2016). However, as previously stated, the year 2000 historical flood caused enormous damage to infrastructure and residents' properties. After the flood event, the local authorities planned and built a cascade of high dykes in order to protect residents as well as crops, which are the major livelihood in this region. In addition, the VMD has great potential with respect to the intensification of agricultural production. In 2009, the area protected by high dykes was about $1,222 \, km^2$, covering around 35 % of An Giang Province; in 2011, this percentage had increased to over 40 % (about $1,431 \, km^2$). Dong Thap has a much lower coverage of about 30 %, corresponding to an area of $990 \, km^2$; however, Dong Thap has deep, inundated areas and its soil contains a high concentration of sulfates, resulting in a low potential for agriculture (Kakonen, 2008).

Figure 4.2 presents the numbers of high dykes and the areas protected by these structures by district in An Giang and Dong Thap provinces until 2011. In 2011, the number of high dykes in An Giang and Dong Thap were 329 and 657 respectively. The total area protected by high dykes in An Giang was larger than in Dong Thap (about 14 compared with 10 km2 respectively). As a result, the mean area of a high dyke in An Giang is larger than in Dong Thap. In fact, high dykes are located mainly along the banks of the Tien River and Hau River (Figure 4.2), where the soils are alluvial (Nguyen et al., 2015).

4.1.3 Flood dynamics in the Mekong Delta

The Mekong Delta is spatially separated into inner and outer sections. The former is dominated by fluvial processes, whereas the latter is dominated by marine processes, including tides and waves (Ta et al., 2002). The Mekong River supplies approximately $416 \, km^3$ of water annually, or 13 200 m3 s−1 through Kratie on average (MRC, 2005). Figure 4.3 shows that water discharge varies from 1700 to 40 000 m3 s−1 between the low-flow and high-flow seasons (Frappart et al., 2006; Le et al., 2007; MRC, 2009b; Wolanski et al., 1996). During the high-flow season, high water discharge causes inundation in the delta floodplains in Cambodia and Vietnam. The annual floods in the Mekong Delta can be indicated by their peaks and volumes. The analysis of flood peaks and volumes at Kratie from 1961 to 2017 shows that the floods in 1991 and 2000 were extreme (Figure 4.4).

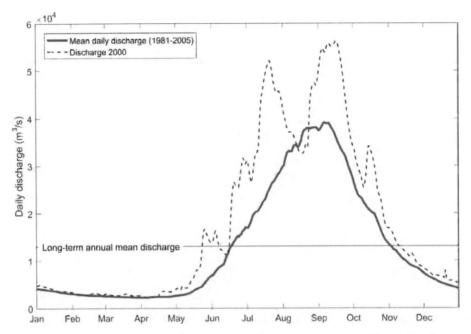

Figure 4.3. Temporal distribution of daily water discharge at Kratie (available data from Darby et al., 2016).

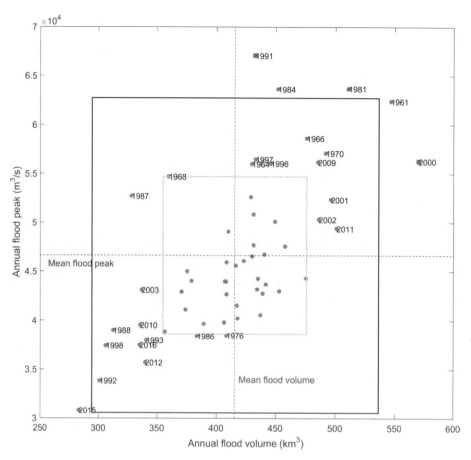

Figure 4.4. The annual flood peaks and volumes at Kratie from 1961 to 2017. The green and black boxes indicate significant (mean ± SD) and extreme (mean ± 2 SD) drought or flood years respectively (SD refers to standard deviation).

From Kratie to Phnom Penh, the hydrodynamics of the Mekong River are dominated by fluvial flows. The river banks are lower than the water levels in the high-flow seasons, which leads to water overflow into the floodplains. The floodplains on the west side convey water to the Tonle Sap River, while the flood water flows into the Tonle Touch River on the east side (Figure 4.1). The floodplains on the west receive less water than those on the east, with water volumes of 24.7 and 35.4 km^3 respectively. The peak discharges of the Mekong River to the respective left and right floodplains are approximately 5,400 and 7,800 m^3 s^{-1} (Fujii et al., 2003). These floodplains in combination with the Tonle Sap River encompass about half of the Mekong's peak discharge.

At Phnom Penh, the Mekong is divided into two branches (the Mekong and the Bassac). In addition the Mekong River joins with the Tonle Sap River. The Tonle Sap Lake is the largest freshwater body in Southeast Asia and has a crucially important role in controlling the water levels in the Mekong Delta. Its surface area covers an area of approximately 3500 km^2 during the low-flow season and is about 4 times larger during the high-flow season (MRC, 2005). The water volume of the lake can reach 70 km^3 in the high-flow season (MRC, 2005). The Tonle Sap Lake functions as a natural flood retention basin for the Mekong River, leading to a reduction in the annual variations of water discharge flowing into the delta. The flood flows to the lake and reverses back to the Mekong River at the Phnom Penh confluence during low flows. Figure 4.5 shows the long-term daily average water discharge flowing in and out of Tonle Sap Lake at the Prek Kdam station. When water levels at Kampong Luong increase, reaching a peak of over 9 m, the lake supplies water to the delta, increasing the Mekong River flows after the flood season and helping to reduce saline intrusion in coastal areas during the low-flow season. From May to September, Mekong water feeds into the Tonle Sap Lake. From October until the following April it then drains back into the Mekong.

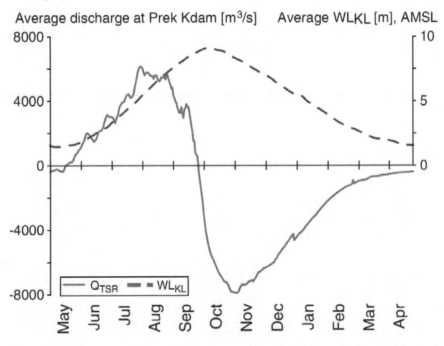

Figure 4.5. Daily averaged (from 1997 to 2004) water discharge hydrograph at Prek Kdam and water level variation at Kampong Luong (Kummu et al., 2014a). The solid line presents the river flow entering Tonle Sap Lake at the Prek Kdam station, and the dashed line shows water levels at the Kampong Luong station.

From Phnom Penh to the Cambodian–Vietnamese (CV) border, the Mekong River flows mainly through the Mekong branch, reaching up to 26 800 m^3/s during flood peaks (Fujii et al., 2003). During these peaks, the floods discharge onto the VMD through the Mekong and Bassac branches as well as via the floodplain overflow comprising 73 %, 7 % and 20 % of the total discharge respectively (Figure 4.1).

In the VMD, the Mekong River flow partly diverts from the Tien River (Mekong branch) to the Hau River (Bassac branch). Regarding flood distribution, water discharge at Tan Chau (Tien River) and Chau Doc (Hau River) is estimated to be 80 % and 20 % of the total flood flow respectively. However, the Vam Nao connecting channel leads to a relative balance between the Tien River (at My Thuan) and the Hau River (at Can Tho) downstream (Figure 4.6). At these stations water levels are strongly dominated by the tides of the East Sea. The water levels in the coastal VMD region fluctuate due to tides from both the East Sea and the West Sea, but the tidal range of the East Sea is much higher than that of the West Sea. Therefore, the East Sea's tides play a more important role and become the main dominant factor controlling hydrodynamics in VMD coastal areas.

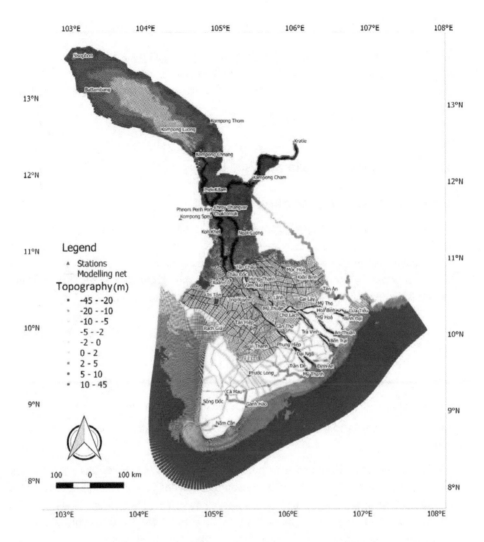

Figure 4.6. Mekong Delta modeling grid and river interpolated topography from 1-D ISIS cross-section data and the shelf topography of the Mekong Delta.

75

4.2 METHODOLOGY

This section introduces the methodology of our study. Section 2.1 describes the model set-up. Section 2.2 provides the model calibration and validation. Section 2.3 and 2.4 elaborate on the high-dyke development scenarios and on further analysis of these scenarios respectively.

4.2.1 Model description and set-up

4.2.1.1 Software description

The hydrodynamic model applied in this study is the Delft3D Flexible Mesh (DFM) Model Suite which has been developed by Deltares (deltares.nl). DFM is a multidimensional model which includes one, two and three dimensions in the same set-up. It solves the 2-D and 3-D shallow water equations (Kernkamp et al., 2011). These equations describe mass and momentum conservation (Deltares, 2020a).

$$\frac{\partial h}{\partial t} + \nabla.(h\boldsymbol{u}) = 0 \ (8)$$

$$\frac{\partial h\boldsymbol{u}}{\partial t} + \nabla.(h\boldsymbol{uu}) = -gh\nabla\zeta + \nabla.\left(vh(\nabla\boldsymbol{u} + \nabla\boldsymbol{u}^T)\right) + \frac{\tau}{\rho} \ (9)$$

where $\nabla = \left(\frac{\partial}{\partial x}, \frac{\partial}{\partial y}\right)^T$, ζ is the water level, h the water depth, u the velocity vector, g the gravitational acceleration, v the viscosity, ρ the water mass density and τ is the bottom friction.

DFM allows computation on unstructured grids, so it is suitable for regions with complex geometry (Achete et al., 2015), including combinations of 1-D, 2-D and 3-D grids. This feature is efficient for taking small canals into account. Therefore, in this study, DFM is selected for simulating flood dynamics in the Mekong Delta which comprises a dense river network and highly variable river widths, dykes and flood plains.

4.2.1.2 Model set-up

The model in this study was improved from the model used by Thanh et al. (2017). In the present configuration, the model uses a depth-averaged setting.

Grid generation and improvement

The unstructured model was constructed using a multi-scale modeling approach; specifically, it consists of a combination of 1-D (canals) and 2-D (the main branches of the Mekong River, its floodplains and shelf) parts. The approach shows efficiency in the case of complex geometry such as the entire Mekong Delta. To capture the hydrodynamics of the main branches and estuaries of the delta, the main channels are represented in enough horizontal detail to resolve the flow patterns over channels and shoals and at the main bifurcations and confluences. Regarding the shelf, the model extended to approximately 80 km from the coastline of the delta to fully contain the river plume (Figure 4.6).

The grid includes the river system of the Mekong River from Kratie to the East Sea and its shelf. The mainstream of the Mekong River, the subaqueous delta and floodplains are represented by 2-D cells, whereas the primary and secondary canals are modelled as 1-D networks. The 2-D cells are a combination of curvilinear (in the main channels) and triangular grid cells. The grid creation was introduced and recommended by Bomers et al. (2019) and Kernkamp et al. (2011). The grid/element sizes vary from about 0.1 km in rivers to 3 km on the delta shelf. The lengths of the grid vary depending on the river geometry. The lengths of cells are generally around 700 m on the Mekong River mainstreams and decrease to approximately 200 m at river bifurcations and confluences. The larger cells of the Tonle Sap Lake, the floodplains and the sea are up to 2000 m. The uniform length of 1-D segments is 400 m. The grid quality is critical for accurate simulations; therefore, the grid has been made orthogonal, smooth and sufficiently dense, to orthogonal values of less than 10 %.

From the survey data, it can occur that a dyke ring in the model can consist of both high dykes and low dykes. This situation may transpire because the model only includes the main rivers and the secondary canal network, and tertiary and small canals are not included. In order to determine whether the dykes are fully protected or partly protected, the ratio of high-dyke area to low-dyke/non-dyke area is calculated. If the ratio is higher than or equal to 1, the dyke is recognised as a high dyke. If the ratio is lower than 1, it is determined not to be a high dyke. In the modeling approach, a high dyke does not allow water flow from linked canals to its protected floodplains, and we did not consider flows over the crest of dykes.

The VMD witnessed three large floods from 2000 to 2002 based on the flood classification of the Tan Chau's flood peaks. Thus, the 2000 and 2001 floods were chosen to calibrate and validate the model respectively. Another reason for selecting the 2000 flood is that the datasets for this flood are comprehensive.

Bathymetry data

When modeling the flood dynamics in the Mekong Delta, bathymetry is a key element. However, available data from the Mekong Delta are limited. For river bathymetry, cross-sectional data were used that were collected by the Mekong River Commission and used to develop the 1-D hydrodynamic model (ISIS) to simulate fluvial flood propagation (Van et al., 2012). To use these profile data for 2-D modeling, the cross-sectional data were interpolated to river bathymetry for the main branches, whereas the primary and secondary canals directly used the cross-sectional data from the 1-D ISIS model. The bathymetry of the sea area is extracted from ETOPO1 (Amante and Eakins, 2009). The floodplains' topography is obtained from the freely available SRTM90 m (Reuter et al., 2007) digital elevation model. Although SRTM is not a high-quality digital elevation model, it was reasonably used for flood modeling in the Mekong Delta (e.g. Dung et al., 2011; Tran et al., 2018).

Boundary conditions

Open boundaries are defined as water discharge (at Kratie) and water levels (the sea). The measured water discharges were used for the upstream boundary at Kratie and were collected from the Mekong River Commission. The latter were defined as astronomical tidal constituents and extracted from a global tidal model (TPXO; Egbert and Erofeeva, 2002). Furthermore, in order to allow for alongshore transport, the northern cross-shore boundary is defined as a Neumann boundary which is driven by the alongshore water level gradient (Tu et al., 2019).

Initial conditions

Water levels in the Mekong Delta vary highly in space due to large-scale flood retention. Thus, the model takes a long time to capture the system behaviour, especially with respect to arriving at the correct flood storage for the Tonle Sap Lake. The Tonle Sap Lake plays a significant role in controlling upstream discharge in the low-flow season. Therefore, the model was spun up over the year 1999; simulated results at the end of this year were used as initial conditions for the year 2000 simulation.

4.2.2 Model calibration and validation

The years 2000 and 2001 were chosen to calibrate and validate the model respectively. The model calibration parameter is the roughness coefficient. This parameter is also selected for calibration without any sensitivity analysis as it is commonly used for calibrating hydrodynamic models (Manh et al., 2014; Wood et al., 2016). In this study, the "trial and error" method is used for calibration. The roughness coefficients are extracted from the previous calibrated models, including ISIS (Van et al., 2012) and MIKE11 (Manh et al., 2014), in order to speed up the calibration process. Model performance in simulating water levels and discharge were evaluated at 36 stations in the Mekong Delta. The model was calibrated against measured data, with the objective function of the Nash–Sutcliffe efficiency (NSE). The NSE is a normalised statistical indicator that uses the comparison of the residual variance and the measured data variance (Nash and Sutcliffe, 1970) and calculated as:

$$E = 1 - \frac{\sum_{t=1}^{T}(Q_m^t - Q_o^t)^2}{\sum_{t=1}^{T}(Q_o^t - \bar{Q}_o)^2} \quad (10)$$

where $\bar{Q}_o$ is the mean of observed discharges, Q_m^t is simulated discharges, and Q_o^t is observed discharge at time t.

In this study, we used different temporal intervals of observation data. Daily data are used in the Cambodian Mekong Delta (CMD), and hourly data are used in the VMD. The reason for this is that hydrodynamics in the CMD are unlikely to be affected by tides, particularly in the high-flow season, whereas hydrodynamics in the VMD are strongly dominated by tides, even in the high-flow season; thus, the hourly data are better for representing tidal fluctuation.

The NSE is commonly used for evaluating hydrological models. Model performance is acceptable if the NSE is higher than 0 (Moriasi et al., 2007). If the NSE is higher than 0, the simulation is a better predictor than the mean observation. A NSE of 1 corresponds to a perfect match between the modelled results and the observed data. The hydrodynamic model is defined as well calibrated if the NSE, in terms of water levels and discharges, is higher than 0.5. Moriasi et al. (2007) classified model performance based on the NSE as "very good" ($NSE > 0.75$), "good" ($0.75 \leq NSE \leq 0.65$), "satisfactory" ($0.65 \leq NSE \leq 0.5$) and "unsatisfactory" ($NSE < 0.5$).

In addition, we used a bias index in order to recognise if the model systematically under- or overestimated water levels. In this study, a commonly used bias measure – the mean error – is used to represent the systematic error of the model (Walther and Moore, 2005). The bias is computed based on the following equation:

$$Bias = \bar{S} - \bar{O} \quad (11)$$

Where $\bar{S}$ is the simulated yearly mean and $\bar{O}$ is the observed yearly mean. The bias is calculated for water levels over the year 2000.

4.2.3 High-dyke development scenarios

To investigate the roles of different floodplains in the VMD and the impact of these floodplains on the VMD's hydrodynamics and downstream tidal propagation, we developed scenarios that include the contributions of each floodplains' water retention. These scenarios used the hydrograph of the year 2000 flood, which was an extremely wet year, in order to estimate the maximum impacts of high dykes. The results of a statistical analysis of flood peaks and volumes encouraged the selection of the year 2000 flood (Figure 4.4).

The hydrodynamic forcing is the same in these scenarios; the only difference is development of high dykes. The scenarios are as follows:

- Scenario 1 (Base) is the base scenario for the 2000 flood, without high dykes. The floodplains in the VMD were not protected by high dykes before 2000 (Duong et al., 2016); therefore, no high dykes are considered in this scenario.

- Scenario 2 (Dyke 2011) includes the high-dyke system in 2011, as illustrated in Figure 4.2. The number of high dykes and the protected floodplain areas are described in Sect. 1.2.

- Scenario 3 (Dyke LXQ) only includes the high-dyke system developed in the LXQ. The floodplain area protected by the high dykes in the LXQ is approximately 3,034 km^2.

- Scenario 4 (Dyke PoR) only includes the high-dyke system developed in the PoR. The PoR is a deeply inundated region in the high-flow season (Kakonen, 2008). In this scenario, the high dykes in PoR protect a floodplain area of around 5,020 km^2.

- Scenario 5 (Dyke Trans-Bassac) only includes the high-dyke system developed in the Trans-Bassac region. This region is a shallowly inundated area comprising 3,152 km^2.

- Scenario 6 (Dyke VMD) assumes that the high-dyke system is totally developed throughout the VMD's floodplains. This scenario is used to investigate the possible impacts of high dykes if they are built to protect the entire VMD floodplain area. The total floodplain area considered in the model is about 13,059 km^2.

4.2.4 Analysis of simulations

4.2.4.1 Tidal harmonic analysis

The peak water level is a good index for indicating extreme events in flooded areas. Tran et al. (2018) and Triet et al. (2017) used the flood peaks to assess the impact of high dykes in the VMD. However, the VMD coastal area is drastically dominated by tides. As a result, the amplitudes of tidal constituents are good indices for presenting average variations in water levels in coastal areas. The water levels at the stations along the Tien River and Hau River were analysed over the whole of the year 2000 using T_TIDE (Pawlowicz et al., 2002).

$$x(t) = b_0 + b_1 t + \sum_{k=1}^{N}\left(a_k e^{i\sigma_k t} + a_{-k} e^{-i\sigma_k t}\right) (12)$$

where N is a number of tidal constituents. We analysed the eight main tidal constituents. Each constituent has a frequency σ_k which is known, and a complex amplitude a_k which is not known. $x(t)$ is a time series. a_k and a_{-k} are complex conjugates.

4.2.4.2 Water balance calculation

To understand flow dynamics, the water balance analysis is conducted using hourly discharge data from simulations. The targeted stations for this analysis are located on the Mekong's mainstreams and boundaries of the flood-prone zones.

$$V_{in}^t = \Sigma_t Q * dt (13)$$

where V_{in}^t is total water volume flowing in the target regions in accordance with the Mekong flow's direction. Q is hourly simulated discharge. dt is the temporal interval, and t is selected periods of an entire year and seasons.

4.3 RESULTS

In this section we present results of model performance and analysis. The model performance with respect to calibration and validation is indicated by the NSE values (Sect. 3.1). The results for the spatial distribution and temporal variation are shown in Sect. 3.2. In addition, Sect. 3.3 presents the impact of high dykes on water levels and tidal propagation.

4.3.1 Model performance evaluation

The overall model performance is generally satisfactory with respect to simulating flood dynamics in the Mekong Delta. For water level calibration, up to 36 stations are used for calibration and the majority of these stations have NSE values higher than the satisfactory level of 0.5. The model performance shows its stability in validation, as the NSE values are higher than 0.7. Generally, the model slightly overestimates water levels. Large biases were found in the CMD, with the largest bias of around 1 m at Kratie. The absolute values of the biases decrease to less than 0.2 m at the stations in the VMD. Particularly, the biases at the middle and coastal VMD stations are smaller than 0.1 m (Table 4.4 in Appendix A).

The annual flood flows through the VMD via the Mekong mainstreams and over floodplains; therefore, discharge data from stations in these areas are employed for calibration. A total of 11 stations on the mainstreams and across the CV border are used for calibration. Simulated and measured discharges at these stations show good agreement, and this is indicated by high NSE values. As a result, the Manning roughness coefficient values of the Mekong River reaches and its floodplains after calibration and validation are illustrated in Table 4.1. The range of roughness coefficients found is relatively similar to previous modeling efforts (Dang et al., 2018a; Manh et al., 2014; Tran et al., 2018; Triet et al., 2017; Van et al., 2012).

Table 4.1. Calibrated values of the Manning roughness coefficient.

River reaches/floodplains	Manning roughness coefficient	River reaches/floodplains	Manning roughness coefficient
Kratie to Phnom Penh	0.031	The Tonle Sap Lake and River	0.032
Cambodia floodplains	0.036	Phnom Penh to Vam Nao (Song Hau)	0.033
Phnom Penh to Tan Chau	0.031	Vam Nao to Can Tho (Song Hau)	0.027
Tan Chau to My Thuan	0.029	Can Tho to Song Hau mouths	0.021
VMD floodplains	0.018	VMD channels	0.027
My Thuan to Song Tien mouths	0.023	Continental shelf	0.016

4.3.2 Spatial distribution and temporal variation of water volume in the VMD

4.3.2.1 Spatial distribution

Water enters the VMD in three ways: the Tien River, the Hau River and flows across the CV border. Figure 4.7 presents the spatial distribution of the water volume in the VMD. The VMD received around 580 km^3 in 2000, with volumes of 405, 83, 61 and 31 km^3 via the Tien River, Hau River, and the right and left CV border respectively. The Tien River diverts a considerable amount (152 km^3) of water to the Hau River via the Vam Nao canal. This is the major mechanism that balances the flows seaward between the Tien River and Hau River. In fact, the streamflows are relatively equal between the Tien River and Hau River, with volumes of 247 (at My Thuan) and 235 km^3 (at Can Tho) respectively. The Tien River is drained by its five estuary branches, whereas the Hau River only has two branches. The Hau River flows into the East Sea discharging 162 and 69 km^3 via the Dinh An and Tran De branches respectively. The Tien River's estuary branches, namely the Cung Hau, Co Chien, Ham Luong, Dai and Tieu, drain a similar volume to the East Sea, with a range of between 54 and 63 km^3, except for the Tieu branch which discharges only 34 km^3.

In addition to the mainstreams of the Mekong River, floodplains have a substantial role in changing the hydrodynamics in the VMD. Hence, we analysed the water balance in the three main flood-prone areas: the LXQ, the PoR and the Trans-Bassac. Among these regions, the PoR harbours the largest amount of floodwater; thus, it is a main flood storage for the VMD. Water primarily flows into the PoR across the eastern part of the CV border in the delta area. In fact, this pathway conveyed approximately 61 km^3 in 2000. The simulated results show a volume deficit of 29 km^3 from the western and southern boundary of the PoR, which is drained to the Tien River. The southern PoR drains a volume of around 38 km^3 to the Soai Rap estuarine branch via the Vamco River. Analysing the water balance of the LXQ shows that it receives water from the northern and eastern sides, while it drains water to the western and southern sides. The yearly inflow to the LXQ is about 44 km^3, with volumes of 31 and 13 km^3 from the northern and eastern boundaries respectively. It is found that a similar amount of water drains out of the LXQ. The LXQ mainly releases water from the western boundary (32 km^3) into the West Sea, followed by the southern boundary (13 km^3). The drained water (11 km^3) from the southern LXQ mostly enters the Trans-Bassac floodplains. An additional source to the Trans-Bassac region is the Hau River, with an annual volume of 6 km^3. The sum of the inflows is drained via the southern canals in this region.

The principal dynamical characteristic of the Mekong Delta floods is their seasonal variation. Figure 4.7 and Figure 4.8 illustrate the seasonal variation of the water volume and the percentage volume (compared with the yearly and seasonal entering volumes at

83

Kratie) respectively. Obviously, the flows in the high-flow season are significantly higher than those in the low-flow season. The flood flows contribute up to between 53 % and 65 % of the annual flows throughout the mainstreams and the percentages increase to over 80 % on the floodplains. The Mekong River that flowed into the VMD in 2000 was about 97 % of the total flow at Kratie. However, the water volume entering the VMD was higher than the entry volume at Kratie in the low-flow season.

In the low-flow season, there are slight discrepancies in the water volumes in the segments of the Mekong River, e.g. from Tan Chau to My Thuan. A part of the discrepant proportion is stored in the river segment. As evidence, the water level at Tan Chau at the beginning of the low-flow season is about 2 m, and it then increases to 3.5 m at the beginning of the high-flow season.

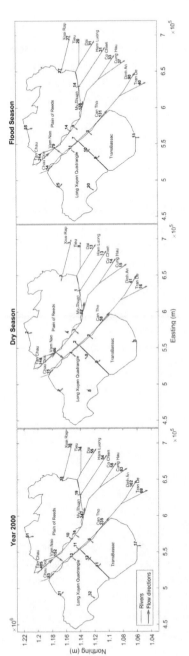

Figure 4.7. The spatial distribution of the water volume (km^3) throughout the VMD in 2000. The low-flow season is calculated from 1 January to 30 June and the high-flow season runs from 1 July to 30 October.

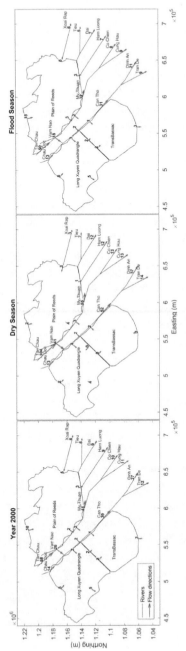

Figure 4.8. Percentages (compared with the total water volume at Kratie) of water distribution throughout the VMD in 2000.

4.3.2.2 Temporal variation

Figure 4.9 presents the simulated fortnightly average discharges at in- and outflows of the Mekong branches as well as cumulative water storage in the main VMD floodplains from April 2000 to April 2001. The yearly average inflows of the Mekong branches at Tan Chau and Chau Doc are approximately 13,000 and 2,700 m³/s respectively. Playing a great role in water diversion between the two Mekong branches, the Vam Nao canal causes water discharges from the Tien River and Hau River to be more balanced seaward of the canal (Figure 4.7). Consequently, the water discharges at My Thuan (Tien River) and Can Tho (Hau River) stations become similar, with annual average amounts of about 7900 and 7500 m³/s respectively. The above-mentioned water at Can Tho station is simultaneously drained through the Hau River mouths. The total outflow from the Tien River is slightly greater than at My Thuan due to added flows from the southern PoR, discharging around 8,400 m³/s.

The water discharges from the Tien River and Hau River are highly variable over time. As shown for the results of discharge variations, the high-flow season is from the beginning of July to the end of October and the remaining period is defined as the low-flow season. The largest seasonal difference is at Tan Chau, with maximum and minimum discharges of about 21,000 and 4,500 m³/s in the high-flow and low-flow seasons respectively. The flood flow at Chau Doc reaches a peak of 5,600 m³/s, while the lowest flow is only 500 m³/s in the low-flow season. However, the high and low flows on the Hau River at Can Tho increase to over 14,100 and 2,200 m³/s respectively. A similar fluctuation is found at the Hau River's mouths. On the Tien River, the flood discharge at My Thuan is only 14,800 m³/s and increases slightly to 17,000 m³/s at the Tien River mouths, but the low flows are similar (2,400 m³/s) at these stations.

The hydrographs in the upstream area of the VMD are flatter than those of the downstream region. The hydrograph shapes are indicated by their kurtosis and are illustrated in Figure 4.9. The kurtosis index is a measure of the peakedness of the distribution. Downstream, the hydrographs are narrower at Can Tho, My Thuan and the outflows of the Tien River and Hau River, with kurtosis values higher than 1.5. One of the noticeable points is that flows at the Can Tho and My Thuan stations are relatively lower at the beginning of the high-flow season than at the end, while the flood flows are stable throughout the high-flow season at Tan Chau and Chau Doc stations. This clearly shows how the early flood peak is stored in the major floodplains of the VMD. Figure 4.9 depicts the cumulative volumes in the major floodplains. At the beginning of the high-flow season, these floodplains are almost empty. By early October, storage increases to 11, 8 and 2 km³ in the PoR, LXQ and Trans-Bassac respectively. When these floodplains are filled, the flood flows at Can Tho and My Thuan reach their maxima during the year.

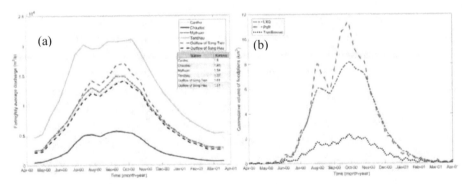

Figure 4.9. Fortnightly average discharges at stations along the Mekong branches (a) and cumulative water volumes of the major floodplains in the VMD (b).

4.3.3 Water level changes under high-dyke development

Figure 4.10 shows that including high dykes increases the daily mean water levels for the Hau River (Chau Doc, Long Xuyen and Can Tho) and the Tien River (Tan Chau, Cao Lanh and My Thuan), especially in the high-flow season. The highest increase was found at Chau Doc and Tan Chau stations while increases decline more seaward.

4.3.3.1 Daily water levels

On the Hau River the floodplains protected by dykes in the LXQ, PoR and Trans-Bassac cause increases of 12.3, 6.1 and 1.1 cm in the annual mean water levels at Chau Doc station respectively. However, Table 4.2 shows that the effect of the PoR dykes on the water levels at Long Xuyen and Can Tho is larger than that of the LXQ dykes. With the high dykes built until 2011, the yearly averaged water levels would increase by 10.2 cm (Chau Doc), 1.5 cm (Long Xuyen) and 0.2 cm (Can Tho). If the high dykes were extended over the VMD (Scenario 6), the yearly mean water levels would increase to 22.3 cm at Chau Doc station.

Generally, water levels on the Tien River are less affected by high dykes. Among the floodplains considered, the PoR has the largest effect on the Tien River's water levels as they are directly connected. For example, the yearly mean water level at Tan Chau increases by about 8.8 cm, but only by 0.6 cm at Cao Lanh station. Interestingly, the PoR slightly reduces water levels at My Thuan due to a reduction in the conveyance capacity of floodwater from the CV border. Although the LXQ is not directly linked to the Tien River, it causes increases in the water levels of around 3.6 and 1.1 at Tan Chau and My Thuan respectively. As the high dykes covered 2421 km² until 2011, the mean water

87

levels are projected to rise by approximately 0.6 cm at My Thuan and up to 6.1 cm at Tan Chau. In addition, the mean water level at Tan Chau could increase by 16.9 cm if the VMD's floodplains were fully protected by dykes. Noticeably, the model errors at these stations were comparable to the variability of water level changes among the scenarios. The differences at the selected stations, except Tan Chau, fell within the model error variations (Table 4.2). The differences in water levels among the scenarios may be influenced by the model set-up.

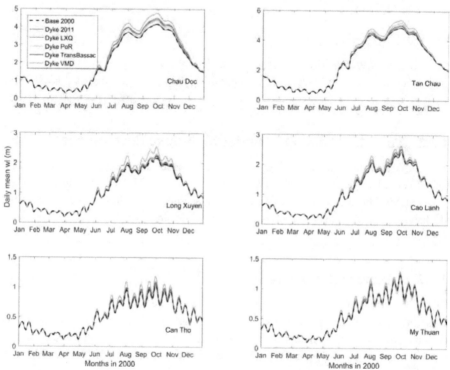

Figure 4.10. Daily mean water level (WL) variations at selected stations along the Hau River (a, b, c) and Tien River (d, e, f) under different high-dyke development scenarios.

Table 4.2. Increases in the annual mean water levels (in cm) over the year 2000 at the selected stations along the Tien River and the Hau River under different high-dyke development scenarios.

Scenario / Station	Song Hau				Song Tien			
	Chau Doc (cm)[a]	Long Xuyen (cm)[a]	Can Tho (cm)[a]	Dinh An (cm)[b]	Tan Chau (cm)	Cao Lanh (cm)[b]	My Thuan (cm)[a]	Ben Trai (cm)[a]
Dyke 2011	10.2	1.5	<u>0.2</u>	0.0	6.1	3.4	0.6	0.0
Dyke LXQ	12.3	3.2	1.0	0.0	3.6	2.6	1.1	0.0
Dyke PoR	6.1	3.6	1.2	0.0	8.8	0.6	-0.8	0.0
Dyke TransBassac	1.1	1.9	0.7	0.0	0.8	0.7	0.3	0.0
Dyke VMD	22.3	9.2	3.0	0.0	16.9	6.6	1.4	0.0

[a] *The differences fall within the model error variations.* [b] *No measured data are available.*

4.3.3.2 Tidal amplitudes

The hydrodynamics in the Mekong Delta are significantly influenced by tides from the East Sea. A tidal harmonic analysis is conducted over the year 2000 to explore possible changes in the main tidal constituents. Figure 4.11 depicts the projected changes in tidal amplitudes along the Tien River and Hau River from the river mouths to approximately 195 km landward under a high-dyke development scenario.

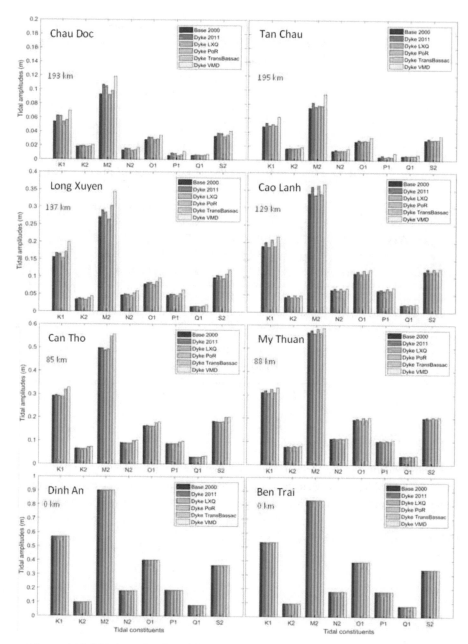

Figure 4.11. Tidal amplitudes of the eight main constituents at the selected stations along the Tien River (right) and Hau River (left) from the river mouths to about 195 km landward in the high-dyke development scenarios.

Tidal amplitudes at the river mouths are unlikely to change. However, differences become significant further inland. At Chau Doc, the LXQ causes the largest increase in tidal amplitudes compared with the other zones. It slightly increases the M2 and K1 tidal amplitude by about 13 % and 15 % respectively. The Trans-Bassac area has a major role in the tidal amplitude change from Long Xuyen to Can Tho. Its dyked floodplains lead to an increase in the tidal amplitudes from 8 % to 13 %. Additionally, the M2 and K1 amplitudes could rise by close to 28 %, 27 % and 12 % at Chau Doc, Long Xuyen and Can Tho respectively. In contrast, high dykes in the PoR result in a marginal reduction in the amplitudes on the Hau River. Similarly, the LXQ and Trans-Bassac cause slight decreases in tidal variation on the Tien River. High dykes constructed on the PoR lead to higher tidal amplitudes on the Tien River, with increases of about 6 %. These increases could reach up to 28 % at Tan Chau, 11 % at Cao Lanh and 4.4 % at My Thuan.

4.4 DISCUSSION

4.4.1 Model performance

The calibration presented in this study considered a larger number of stations compared with previous studies (e.g. Thanh et al., 2017; Tran et al., 2018; Triet et al., 2017; Van et al., 2012). These stations are mainly located in the VMD (Figure 4.13). The majority of stations on the branches of the main Mekong River have a NSE higher than 0.8 (good). In contrast, the stations located further away from the main stream have a lower NSE values. The NSE values of the Phuoc Long and Ca Mau stations are lower than the acceptable level because the water levels at these stations are highly influenced by local infrastructure, specifically the Quan Lo Phung Hiep (QLPH, see Figure 4.1) project. The QLPH has been constructed to protect this area from saline intrusion. Flows entering the QLPH are controlled by a series of sluice gates that are mainly located along the coast to prevent saline intrusion into areas of rice cultivation and to control fresh water sources. We did not consider these sluice gates in the model, as they do not have a fixed operation schedule: their operation is based on crop calendars and in-situ hydrodynamics (Manh et al., 2014). For example, the observed water level at Phuoc Long station is relatively unchanged at 0.2 m over the year 2000, while the model estimates that water levels at this station have semi-diurnal variations between −0.2 m and 0.6 m due to the tidal effects from the East Sea. During validation, a better fit was found at My Thuan station, whereas the other stations have comparable NSE values. As such, we are confident that the model is capable of capturing hydrodynamics in the Mekong Delta accurately.

The modeling approach in this study overcomes a limitation of previous 1-D models that define their boundaries at the river mouths. The boundary conditions (usually water levels) at these locations are not always available, as the water level measurement system in the VMD is not installed at all river mouth locations. Imposing a simple tidal forcing is not justifiable because river flow will impact the mean water level as well as the tidal characteristics in river mouths. Our model grid, which considers part of the shelf, allows for a proper description of these dynamics.

4.4.2 Spatiotemporal distribution of water volume in the VMD

The total net water volume flow through the Mekong Delta at Kratie was approximately 600 km^3 in 2000, as the annual flood contributed about 480 km^3 at this location. This is considerably higher than the average volume of 330 km^3. However, the annual flood peak in 2000 is only slightly higher than the mean flood peak of 52 000 m^3 s^{-1} (MRC, 2009a). Thus, the 2000 flood is characterised by a broader than usual hydrograph.

Several studies have investigated the distribution of the flood volume in the Mekong Delta (e.g. Manh et al., 2014; Nguyen et al., 2008; Renaud and Kuenzer, 2012). Manh et al. (2014) calculated the flood volume distribution for the floods between 2009 and 2011 in the upper VMD and concluded that the flood distribution changed marginally over the above-mentioned period. However, they did not estimate flow distribution through the river mouths. We found a similar pattern with respect to the flood volume distribution on the mainstreams, but our model estimated a larger discharge across the VC border to the VMD. A possible explanation for this is that the 2000 flood was considerably larger than the floods during the 2009–2011 period. Table 4.3 shows a comparison of the VMD's outflows from the current study and from five other models, as summarised by Nguyen et al. (2008). There is only a small variation among the models used which is attributed to different topographical data and boundary conditions (Nguyen et al., 2008). The flow distribution from the current study falls within the range of variation of the other five models, although it differs regarding some important branches, such as the Tien River and Hau River, below Vam Nao.

Table 4.3. Distribution of water discharge throughout the river mouths (following Nguyen et al. (2008).

Model name	The Song Tien below Vam Nao (%)	The Song Hau below Vam Nao (%)	Co Chien (%)	Cung Hau (%)	Dinh An (%)	Tran De (%)	Ba Lai (%)	Ham Luong (%)	Tieu (%)	Dai (%)	Others (%)
NEDECO 1974	51	49	13	15	28	21	0	15	2	6	0
VNHS 1984	55	45	13	18	27	18	0	17	1	6	0
SALO89 1991	44	54	12	8	26	24	2	14	5	2	8
Nguyen Van So 1992	–	–	11	12	19	16	1	14	1.5	6	20
VRSAP 1993	50	44	11	5	18	18	0	9	2	8	29
This study	41	39	10	11	27	12	0	9	6	9	14

Note that the percentages in this study are calculated based on the total volume at Kratie.

The water distributions slightly vary over the high-flow and low-flow seasons. The largest changes are found in the discharges to the floodplains. For instance, water volumes are highly seasonal at the CV border stations. The water flows in the low-flow season contribute to 2 %–6 % of the annual flows at these stations. The relative percentages of the Mekong flow, exiting via the Hau River estuaries in the low-flow season are higher than those in the high-flow season, whereas the percentages at the Tien River estuaries are relatively constant.

Several studies have investigated the roles of the Tonle Sap Lake in regulating the flood regimes on the Mekong River (Fujii et al., 2003; Kummu et al., 2014a; Manh et al., 2014). Kummu et al. (2014a) estimated that the Tonle Sap Lake is capable of reducing about 20 % of the Mekong mainstream discharge and that its greatest storage volume is in August, with an amount of around 15 km^3 from the Mekong River flows. The highest monthly released volume occurs in November and peaks at nearly 20 km^3. Consequently, Tonle Sap Lake has a crucial role in regulating the Mekong River flows on a temporal scale. The VMD floodplains have a different role than the Tonle Sap Lake with respect to changing the Mekong mainstream flows. They mainly store early flood waters in August. This leads to reduce flood flows at downstream stations along these floodplains. These stations reach peak discharges when the VMD floodplains are almost fully filled. Therefore, the peak flows at the downstream stations occur in October. These results are consistent with the analysis of Dang et al. (2018b).

4.4.3 Impact of high-dyke development

The Mekong Delta is presently facing several threats, such as the impact of hydropower dams, sea level rise, delta land subsidence and hydraulic infrastructure (Dang et al., 2018a; Kondolf et al., 2018). The impacts of these threats are highly various in terms of their timescales. Among these, hydraulic infrastructure (especially high dykes) has a considerable influence on the hydrodynamics in the region on a short timescale. The high dykes in the VMD are built to protect agricultural land during floods. As a result, flood discharges on the rivers increase and hydrodynamics in the VMD change. Specifically, the results indicate that a lack of flood retention in the LXQ leads to an increase in the water levels on the Hau River, with a downward trend of increases from Chau Doc to Can Tho. This rising pattern was also found by Tran et al. (2018), albeit with different magnitudes due to the different years. They compared the peak water levels, whereas we used daily mean water levels for comparison. Tran et al. (2018) found that the water level peaks would be drastically higher if high dykes were built. These peaks especially increased in the upper VMD (e.g. by 66 cm at Chau Doc and only 4 cm at Can Tho).

Interestingly, the high dykes in the PoR have slightly stronger impacts on water levels at Long Xuyen station than those in the LXQ region. The reason for this is an increase in the water levels on the Tien River, which causes an increase in the water diversion from the Tien River to the Hau River. Due to the Vam Nao connecting canal, the PoR floodplains influence water level fluctuation not only on the Tien River, but also on the Hau River. In addition, the LXQ floodplains affect water levels on both the Tien River and Hau River. Nevertheless, the increasing water levels on the Tien River remain slightly lower than the levels on the Hau River, as the Tien River has more river mouths and a higher conveyance capacity in comparison with the Hau River.

Recent studies on the impact of high dykes in the VMD (e.g. Tran et al., 2018; Triet et al., 2017) have only compared the maximum water levels. However, we found that high dykes also resulted in a reduction in the minimum water levels. This means that high dykes have effects on tidal fluctuation on the main branches. We analysed the tidal amplitudes of the eight main constituents over the year 2000 in order to quantify how water levels on the main branches changed. Noticeably, the complete implementation of the high-dyke system over the VMD floodplains can cause increases of about 12 % and 4 % in the tidal amplitudes at Can Tho and My Thuan stations respectively. Additionally, high dykes in the PoR directly adjacent to Tien River cause a reduction in the tidal amplitude on the Hau River and vice versa. The reason for this is that river water cannot flow into the floodplains, which leads to an increase in river discharge in the main streams. This increased river discharge causes a significant M2 amplitude reduction (Guo et al., 2016). The amplitudes and mean water level at the river mouth stations are unlikely to change under high-dyke development (Table 4.2). This is due to the fact that flood retention loss due to floodplain areas protected by high dykes causes an insignificant change in the water discharge at that location. In contrast, and as an example, Kuang et al. (2017) found that if water discharges from the Yangtze River upstream increased by $20\,000\,\mathrm{m^3\,s^{-1}}$, the water levels at the mouth could rise by approximately 1 cm. An explanation for this is that the water discharge change due to high-dyke development is not large enough to increase water levels at the river mouths.

The impacts of high-dyke development on the downstream hydrodynamics are considered to be different and not as significant as those of hydropower dams, climate change and sea level rise. Dang et al. (2018a) revealed that hydropower development increases the monthly water levels at Tan Chau by 0.4 m and at My Thuan by 0.05 m in a wet year. Sea level rise causes a gradual but drastic increase in water levels. For instance, the water level at My Thuan could increase by 0.3 m due to a sea level rise of 0.38 m. However, high-dyke development in the VMD causes another distinct effect on the floodplains: it prevents flood waters from entering the floodplains, thereby excluding sediment deposition on these floodplains. Sediment deposition on the floodplains benefits agricultural production. Chapman et al. (2016) estimated that annual sediment deposition in An Giang would be worth USD 26 million.

4.4.4 Flood discharge and volume scenarios

In order to investigate the impact of the distribution of hydrological conditions, we selected 1981, 1991 and 2000 for comparison, as these years include extreme floods in terms of the flood peaks and highest volumes. The simulations considered conditions with and without high dykes. Figure 4.12 shows that the water levels at the VMD stations are comparable for all 3 years, although the 1991 peak flood at Kratie is about 20 % larger than the 2000 peak flood. Furthermore, the year 2000 flood – which had the largest flood volume – led to the highest water levels. Thus, flood volume is more important than flood peak flows for extreme flood conditions. The reason for this is that high flood peaks flood the area and fill the Tonle Sap Lake in Cambodia; thus, the peak flood flow decreases and elongates downstream (Triet et al., 2017). In addition, Figure 4.12 shows that the water level difference for scenarios with and without high dykes is similar and decreases downstream for all 3 years.

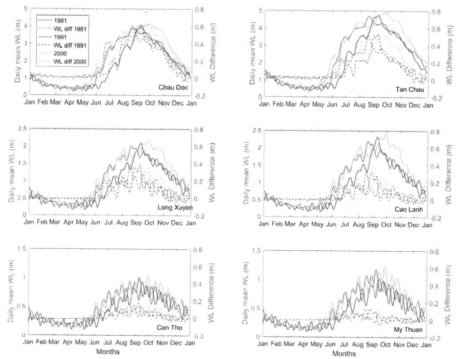

Figure 4.12. Simulated daily mean water levels in the high-flood years of 1981, 1991 and 2000 at selected stations on the Mekong River. The dashed lines indicate water level differences due to the impacts of high dykes.

4.4.5 Estimated uncertainties

The quality of input data directly influences the uncertainty of the model results. There are varying sources of uncertainty such as topography, river discharge, initial condition and model parameters (Abily et al., 2016; Di Baldassarre and Montanari, 2009; Bates et al., 2004; Savage et al., 2016; Teng et al., 2017). The river discharge data contain uncertainty due to the use of a rating curve to generate a discharge value from the water level. The interpolated and extrapolated discharge have been shown to be uncertain by between 6.2 % and 42.8 % (Di Baldassarre and Montanari, 2009). The initial condition includes water levels as the initial state of the rivers and floodplains, especially for the Tonle Sap Lake. However, there is a general lack of measured water levels throughout the entire Mekong Delta. To overcome this limitation, we adopted a long spin-up time. River bathymetry in this study was reconstructed from cross sections. Although the river bathymetry was reconstructed using an efficient interpolation method, it contains an error of 0.74 m (Thanh et al., 2020b). One of the major sources of uncertainty stems from the SRTM data. Although the SRTM data were used for the VMD floodplains, they contribute to the model error. Recently, several studies have reported that the SRTM data contain high uncertainties, including stripe noise, speckle noise, absolute bias and tree height bias (Hawker et al., 2018; Tarekegn and Sayama, 2013; Yamazaki et al., 2017). In fact, Hawker et al. (2018) revealed that the SRTM data could contain vertical height errors ranging from 1.0 to 4.8 m. These studies suggest that the SRTM data should be processed to remove these errors. However, the model in this study enables one to better understand the hydrodynamics in the Mekong Delta and might serve as a tool for comparative studies. Future work will explore the impact of model input uncertainties on the model outcomes.

4.5 CONCLUSIONS

In this study, we applied a process-based model (DFM) in order to simulate hydrodynamics in the entire Mekong Delta from Kratie to the shelf areas. The model was calibrated using a dataset of water levels and discharge at 36 stations over the Mekong Delta. The model shows good agreement between simulations and observations. This model is an improved version of the model used by Thanh et al. (2017), as it takes the Cambodian and Vietnamese floodplains as well as the dense river/canal network in the VMD into account. Nevertheless, it does not contain tertiary rivers/canals and hydraulic structures for salinity regulation.

We found that the change in seasonal flow distribution throughout the Mekong's mouths is insignificant, except at the Dinh An mouth which shows a slight increase in the low-flow season. In contrast, the Mekong River network discharging to the sea via the Soai Rap mouth and the western LXQ dramatically dropped in the low-flow season compared with the high-flow season due to overflow reduction at the CV border.

This study found that the floodplains protected by dykes in the LXQ and PoR influence water regimes not only on the directly linked Mekong branch, but also on other branches. The LXQ high dykes cause an increase in the daily mean water levels, but a decrease in tidal amplitudes on the Tien River (after the Vam Nao connecting channel). A similar pattern is also found for the interaction between the PoR high dykes and the Hau River.

The high dykes built in the PoR, LXQ and Trans-Bassac regions have demonstrated an impact on water levels at Tan Chau, Chau Doc and Can Tho respectively. These outcomes will benefit sustainable water management and planning in the VMD.

APPENDIX: MODEL CALIBRATION

For water level calibration, up to 36 stations are used for calibration. Based on the above-mentioned classification, the NSE values at 33 of 36 stations are higher than the acceptable value of 0.5: 23, 4 and 6 stations are classified as very good, good and satisfactory respectively (Figure 4.13). The three unsatisfactory stations are at Ca Mau, Phuoc Long and Rach Gia. Regarding the model validation, only 14 stations are used due to the availability of observed data. Over the validation period, these stations have NSE values that are in the same groups as the calibration values. However, while they are in the good and very good classes, the values at the My Thuan stations increase from 0.69 in calibration to 0.74 in validation.

For water discharge calibration, 11 stations on the mainstreams and across the CV border are used for calibration. A total of 9 of the 11 stations have NSE values in the very good category (Figure 4.14). The two stations cross-border stations, namely the right border (to the PoR) and left border (to the LXQ), are in the satisfactory and unsatisfactory categories respectively; however, the NSE values are 0.49 at the left border and 0.54 at the right border, fluctuating around the acceptable criteria. Data at these stations are not available for validation. All stations used for validation are in the very good group. Compared with the calibration, the NSE values for the validation are relatively stable, except for My Thuan station, which increases from 0.84 to 0.95.

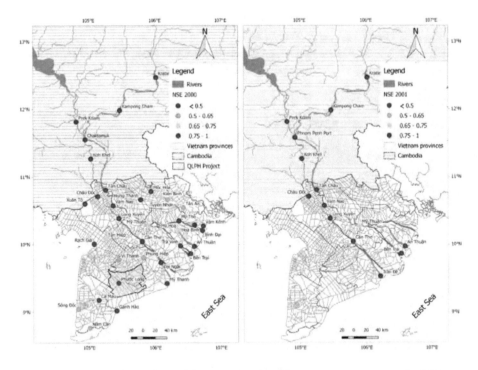

Figure 4.13. The NSE values of the water levels at the gauging stations in the Mekong Delta. The calibration (the year 2000) and validation (the year 2001) are presented in panels (a) and (b) respectively.

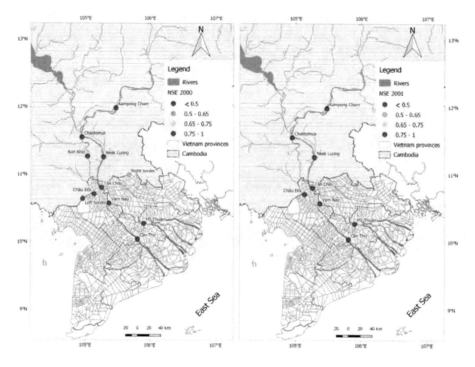

Figure 4.14. The NSE values of the water discharges at the gauging stations in the Mekong Delta. The calibration (the year 2000) and validation (the year 2001) are presented in panels (a) and (b) respectively.

Table 4.4. Calculated bias for water level calibration at stations in different regions.

CMD	Bias (m)	Upstream VMD	Bias (m)	Middle VMD	Bias (m)	Coastal VMD	Bias (m)	Coastal VMD	Bias (m)
Kratie	1.02	Tan Chau	0.12	Kien Binh	−0.03	An Thuan	−0.02	Nam Can	−0.24
Kampong Cham	0.54	Chau Doc	−0.20	My Thuan	−0.08	Ben Trai	−0.04	Phung Hiep	−0.07
Prek Kdam	0.92	Vam Nao	−0.10	Can Tho	0.02	Binh Dai	0.00	Phuoc Long	−0.06
Phnom Penh	−0.15	Xuan To	0.09	Hung Thanh	0.00	Ca Mau	−0.18	Rach Gia	−0.08
Koh Khel	−0.51	Moc Hoa	0.04	Tuyen Nhon	0.06	Dai Ngai	−0.02	Song Doc	−0.03
		Long Xuyen	−0.12			Ganh Hao	−0.07	Tan An	0.08
						Hoa Binh	0.09	Tan Hiep	−0.10
						My Hoa	0.00	Tra Vinh	0.05
						My Thanh	0.02	Vam Kenh	0.07
						My Tho	0.02	Vi Thanh	0.00

5

A NUMERICAL INVESTIGATION ON THE SUSPENDED SEDIMENT DYNAMICS AND SEDIMENT BUDGET IN THE MEKONG DELTA[4]

Abstract

Fluvial sediment supply towards the coast has been the subject of extensive research. Important aspects relate to the impact of sediment retaining hydropower dams, potential delta progradation, coastal sediment supply and delta vulnerability to sea level rise. Once validated, process-based models provide a valuable tool to address these aspects and offer detailed information on sediment pathways, distribution and budget in specific systems. This study aims to advance the understanding of the sediment dynamics and sediment budget in the Mekong Delta system. We developed a process-based model (Delft3D FM) that allows for coupling 2D area grids to 1D network grids. The flexible mesh describes both wide river sections and channel irrigation and drainage networks present in the Mekong Delta. We calibrated the model against observed discharge, salinity, suspended sediment concentration (SSC) and sediment flux. The model was able to skillfully describe seasonal variations of SSC and hysteresis of SSC and water discharge. The Tonle Sap Lake is a major cause of the concentration and discharge hysteresis, which results from seasonal bed sediment availability in the channels. Model results suggest that the

---

[4] This chapter is based on:

Thanh, V. Q., Roelvink, D., van der Wegen, M., Reyns, J., van der Spek, A., Vinh, G.V. and Linh, V.T.P.: A numerical investigation on the suspended sediment dynamics and sediment budget in the Mekong Delta, Cont. Shelf Res., 2021 (under review).

Mekong River delivers ~99 Mt of sediment towards its delta through Kratie, which is much lower than the common estimate of 160 Mt/year. About 23% of the modelled total sediment load at Kratie reaches the sea. Our modeling approach is a useful tool to assess sediment dynamics under strategic anthropogenic interventions or sea level rise.

5.1 INTRODUCTION

The worldwide fluvial sediment flux to coastal deltas amounts to 12.8 - 15.1 Gt per year (Syvitski and Kettner, 2011). Understanding sedimentary processes in these deltas is important to estimate the impact of anthropogenic strategies for sustainable management and to address the impact of climate change like changing river flow, sediment supply and sea level rise. Sediment fluxes are commonly estimated in relation to river discharge (Ogston et al., 2017), but this is associated with high uncertainties due to sparse data, both in time and space. Tidal influence makes sediment dynamics in deltas even more complex to understand, while local conditions make every Delta unique.

The Mekong Delta is crucial to the local livelihoods and food security. The area is home to about 17 million people in the VMD (Vo, 2012). Particularly, the Vietnamese Mekong Delta (VMD) is a "rice bowl" for Vietnam and for the world. The VMD covers a region of 39,700 km^2 and ~60% of this area is for agricultural cultivation (Vo, 2012). One of the important factors favouring agricultural cultivation is abundant availability of water and sediment from the Mekong River. Annually, the Mekong River supplies ~416 km^3 of water and delivers ~73 Mt sediment towards its Delta (Koehnken, 2014; MRC, 2005; Thanh et al., 2020a). The annual sediment transport at Kratie (Cambodia) varies in a range of 44 – 98 Mt/y from 2009 to 2013. These values fell within long-term estimates (87.4 ± 28.7 Mt/y) of Darby et al. (2016). Recent studies (e.g. Darby et al., 2016; Kummu and Varis, 2007; Lu et al., 2014; Manh et al., 2014) confirmed that the suspended sediment flux into the Mekong Delta (at Kratie station) is much less than the commonly used value of 160 Mt/y. However, prior to the construction of hydropower dams on the Mekong River mainstream, sediment loads could have been substantially higher than afterwards. The hydropower dams not only change the seasonal flows but also store sediment in their reservoirs. Lauri et al. (2012) found that these reservoirs can increase flows at Kratie in the low flow season by 25-160% and decrease flows in the high flow season by 24%. Annual floods are the main source of fresh water in the VMD while the sediments delivered act as a natural and valuable fertilizer source for agricultural crops (Chapman and Darby, 2016). However, the VMD is facing challenges related to flood regimes and sediment transport due to climate change, sea level rise and human interventions.

Sediment transport in the Mekong River has been estimated by *in-situ* measurements, sediment rating curve methods; and numerical modeling. Sediment measurements in the Mekong started in the 1960s, inspired by US practices (Walling, 2009). When using data-based methods, the reliability of sediment transport depends on the number of measuring stations, the length of record and temporal resolution of the data. It has been hardly possible to cover a large area like the Mekong Delta with measurements. In addition, discontinuous records and low sampling frequency lead to high uncertainties in sediment budget estimations. A numerical model, calibrated by *in-situ* measurements and rating curves, is a suitable tool to investigate hydrodynamics and sediment transport in the Mekong Delta in more detail.

There is a large number of studies focusing on sediment dynamics in the VMD, ranging from measurement based studies to numerical modeling studies. Hung et al. (2014), Manh et al. (2013) and Nowacki et al. (2015) provide *in-situ* measurements on a limited amount of locations in the VMD. The data of *in-situ* measurement are accurate, but it is difficult to collect them on a large spatial scale to derive sediment budgets on the scale of the entire VMD. However, *in-situ* measurement data are essential to calibrate and validate numerical models.

Numerical models for the entire Mekong Delta are commonly set up using a 1D schematization (Manh et al., 2014), while smaller-scale area models are represented by 2DH or 3D setups (Marchesiello et al., 2019; Thanh et al., 2017; Tu et al., 2019; Xing et al., 2017). For example, Thanh et al. (2017) and Tu et al. (2019) used a 3D, process-based model (Delft3D4) to investigate sediment dynamics and morphological changes in the coastal area of the Mekong Delta. With a similar approach, Xing et al. (2017) developed a model for the lower Song Hau channel to advance the understanding of hydrodynamics and sand transport in this region. Both the studies of Thanh et al. and Xing et al. used two spatial scales for modeling, including a large and coarse grid to act as boundary condition for a fine and detailed grid. The use of this approach can reduce computational cost, but it may cause significant uncertainties. Therefore, creating a single model domain for the entire Mekong Delta and its shelf could result in accurate results. A large part of the Mekong Delta consists of a dense channel network, with high variability in channel widths, which can be approached by a 1D network. A pure 2D model for the entire Mekong Delta would be unnecessary and computationally inefficient. 3D effects like gravitational circulation could be relevant only in more seaward reaches. In this study, we propose a 1D-2D coupled model for the Mekong Delta and shelf. The main channels of the Mekong River and floodplains are modelled in 2D while primary and secondary channels are represented by 1D elements. This approach was efficient for large scale and complex regions and it accurately modelled hydrodynamics in the whole Mekong Delta (Thanh et al., 2020a).

The objective of this paper is to derive a sediment budget for the Vietnamese Mekong Delta for the high river flow year of 2011 using a 1D-2D coupled, process-based model (Delft3D Flexible Mesh, DFM). In section 5.2 we will introduce the Mekong Delta and its sediment characteristics. In section 5.3 we first describe the model DFM and the modeling approach for the Mekong Delta. In section 4, the model calibration is presented and we investigate sediment dynamics and estimate a sediment budget in the Mekong Delta. Finally, section 5.5 presents conclusions.

5.2 CASE STUDY DESCRIPTION: THE MEKONG DELTA

5.2.1 Characterization of the Mekong Delta

The Mekong Delta is the third largest deltas in the world (Anthony et al., 2015). It has been formed over 6,000 years ago in response to decelerating sea level rise (MRC, 2010). It covers an area of about 60,000 km^2 and is home to about 17 million people (Vo, 2012).. The Mekong River is one of the world's largest rivers, with a length of approximately 4,800 km and its draining catchment area of 795,000 km^2 (MRC, 2005). It flows through the six countries, originating from China, through Myanmar, Laos, Thailand, Cambodia and Vietnam, before debouching into the East Sea (South China Sea). The Mekong Delta is commonly defined from Phnom Penh downstream, where the Mekong river is separated into two branches, namely Mekong and Bassac (Gupta and Liew, 2007; Renaud et al., 2013). This area is partly located in Cambodia and Vietnam. The Mekong Delta in Cambodia (CMD) and in Vietnam (VMD) have different hydrological regimes. An important confluence of the Mekong River and the Tonle Sap River, located at Phnom Penh, is responsible for this. During the initial phase of annual floods (July - October), the Mekong River also fills the Tonle Sap Lake via the Tonle Sap River. At decreasing flood flows, the lake empties again via the Tonle Sap River into the Mekong River. The lake thus lowers and elongates yearly hydrographs. In order to understand hydrodynamics and sediment dynamics in the Mekong Delta, extending the area up to Kratie (Figure 5.1) is needed (Thanh et al., 2017).

The CMD encompasses a large area of lowland which is deeply inundated by the annual floods. For instance, inundation depths on the CMD floodplains are generally over 3 m (Fujii et al., 2003). The hydrodynamics of the Mekong River in the CMD is dominated by the annual floods which are considerably changed due to the southwest monsoon (Yu et al., 2018). In addition, the hydrodynamics in this region are also influenced by regulation of the Tonle Sap Lake.

The VMD covers an approximate area of 40,000 km^2 and is home to about 17 million people (Vo, 2012). It has a complex river network which contains a large number of man-made canals Extensive   canal development for agricultural purposes started in 1819 (Hung, 2011). Seventy-five percent of the VMD area is used for agricultural production (Kakonen, 2008). Recently, several hydraulic structures have been constructed in the VMD to protect agricultural crops, such as dyke rings, sluice gates and culverts. These modifications have considerably changed the hydrodynamics in the VMD (Thanh et al., 2020a; Tran et al., 2018).

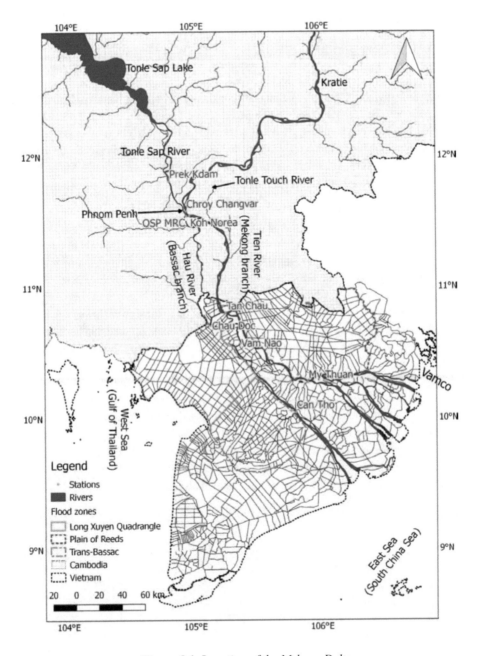

Figure 5.1. Location of the Mekong Delta.

5.2.2 Sediment loads

There are two types of sediment loads towards the Mekong Delta. Suspended sediment loads at Kratie occupy 97% of the total sediment load while the bedload is only 3% (Koehnken, 2012). Therefore, we focus on the suspended sediment load in this study. Milliman and Syvitski, 1992 and Walling, 2008 estimate the annual sediment load of the Mekong River to be 160 Mt/y. Sediment loads at Kratie fluctuated between 23-134 Mt/y from 1982 to 2004 with an average load of approximately 87 Mt/y (Darby et al., 2016), including extremely wet years (e.g. 2000). Koehnken (2014) estimated the annually averaged sediment load at Kratie from 2009 to 2013 slightly lower at about 73 Mt/y while the annual sediment load varied between 44 and 98 Mt/y in 2010 and 2011 respectively.

The Mekong River is subject to strong seasonal fluctuations and sediment loads vary accordingly. In general, sediment loads at Kratie in the high flow season (from July to October) provide approximately 95% of the annual sediment loads (Dang et al., 2018b; Koehnken, 2014). The greatest sediment load usually occurs in September, supplying 25-40% of the annual load. In contrast, monthly sediment loads in the low flow seasons are extremely small, with a contribution of < 1% of the annual load (Koehnken, 2014). From Kratie downstream, sediment loads are spatially correlated to local river flows in general, except for Tonle Sap River's sediment loads. Before the Mekong-Tonle Sap confluence, sediment loads at Chroy Chang Var (Figure 5.1) are comparable to those at Kratie, with a highest suspended sediment load of about 1.4 Mt/day (Figure 5.2). At the Tonle Sap-Mekong confluence, most sediment is transported to the Mekong branch via Koh Norea station, while the amount of sediment transported through the Bassac branch at station OSP MRC is much smaller, with a ratio of 1/6. The sediment flux into the Tonle Sap River mainly occurs during the early flood stage. The annual inflow into and outflow from the Tonle Sap River is about 6.4 Mt and 1.5 Mt (Koehnken, 2012). This ratio is consistent with model results computed by Kummu et al. (2008). The difference indicates the sediment trapping efficiency of Tonle Sap Lake, of around 80% (Sarkkula et al., 2010).

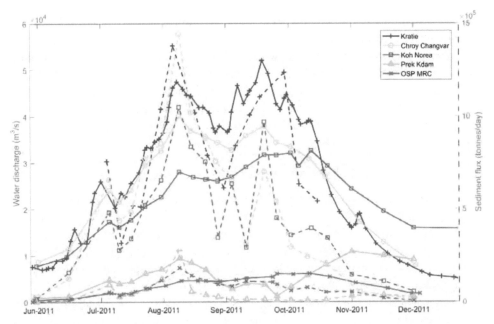

Figure 5.2. Suspended sediment fluxes and water discharge on the main channels of the Mekong River in 2011 (aggregated from Koehnken, 2012). The dashed lines present sediment flux while the hatched lines show water discharge at the selected stations.

The Mekong (Song Tien) and Bassac (Song Hau) branches both supply suspended sediments to the VMD. The total sediment loads in 2011 were 50 Mt at Tan Chau on the Song Tien and 9 Mt at Chau Doc on the Song Hau (Manh et al., 2014). The connecting channel of Vam Nao diverted an amount of around 19 Mt from the Song Tien to the Song Hau in 2011 and balanced the sediment fluxes of the Song Tien and the Song Hau. As a result, sediment fluxes at My Thuan on the Song Tien and Can Tho on the Song Hau were approximately 26 and 29 Mt/y in 2011 (Manh et al., 2014). Nowacki et al. (2015) estimated that the Song Hau and the Song Tien mouths exported sediment amounts of 15 and 25 Mt/y in 2012-2013, respectively.

5.2.3 Suspended-sediment concentration

The suspended sediment concentration (SSC) in the Mekong Delta is typically smaller than 0.5 g/l (Koehnken, 2012; Manh et al., 2014), and strongly modulated by the annual floods. In the Cambodian Mekong Delta, SSC in the Mekong River main tributaries fluctuates between 0.2-0.4 g/l during high flow seasons. The SSC on the Tonle Sap River is smaller than that on the Mekong River, with concentrations below 0.2 g/l (Koehnken,

2012). In the VMD, the hydrodynamics are not only influenced by the annual floods, but also by tides. At Can Tho station, the monthly average SSC is about 0.05 g/l, and it can increase to 0.18 g/l in the high flow seasons and decrease to 0.03 g/l in the low flow seasons. The SSC at ebb tides is slightly higher than at flood tides in the low flow seasons. This discrepancy increases in the high flow seasons (Dang et al., 2018b). A similar pattern is found at My Thuan station, with slightly higher SSC. SSC near the Dinh An mouth was low, smaller than 0.03 g/l, in the high flow season (Nowacki et al., 2015).

5.2.4 Sediment grain size distribution

In general, suspended sediment grain-sizes vary seasonally and spatially in the Mekong Delta. Grain-sizes of suspended sediment spatially decrease with distance downstream. Koehnken (2014) reported on a large-scale sediment monitoring campaign in the lower Mekong River. They found predominant cohesive sediments at Kratie. A small amount of fine sand was detected during high flow seasons. The suspended sediment load at Kratie comprises 20% of sand, 61% of silt and 19% of clay materials. Finer sediments were found at Tan Chau, with proportions of 1%, 44% and 54% for sand, silt and clay respectively (Koehnken, 2014). Sarkkula et al. (2010) found that d_{50} is only 3-8 μm at Tonle Sap River and even finer in the Tonle Sap Lake. Hung et al. (2014) carried out an in-situ measurement of sedimentation in the upper VMD. Their results show that that d_{50} is from 10 to 15 μm on the Plain of Reeds (PoR)'s floodplains. (Wolanski et al., 1996) measured that d_{50} is from 2.5 to 3.9 μm in the freshwater regions of the Song Hau estuarine branch. An important sediment process in the estuarine reaches is flocculation that leads to flocs much larger than the individual grain sizes. For example, Wolanski et al. (1996) found that d_{50} of a floc is around 40 μm at the Song Hau estuary. This is consistent with the results presented by Mclachlan et al. (2017) who show that the typical recorded sediment grain size is about 40 μm in the Song Hau estuary. Moreover, this size of flocs is similar to the typical sediment grain size found in the PoR's floodplains (Hung et al., 2014b).

5.3 METHODOLOGY

5.3.1 Software description and model setup

5.3.1.1 Description of Delft3D FM

Hydrodynamics and sediment transport are modelled by flow and sediment transport modules which are combined in the Delft3D FM (DFM) modeling suite developed by Deltares (Deltares, 2020a). The DFM is the successor of Delft3D4 which has been widely used for hydrodynamic modeling of seas, rivers and floodplains. DFM noticeable improvement is the use of unstructured grids and concurrent multi-dimensional modeling, encompassing 1D, 2D and 3D domains. Achete et al. (2016), Martyr-Koller et al. (2017) and Thanh et al. (2020a) provide examples of a successful 2D and 3D DFM model descriptions and validation in estuarine environments.

The flow module of DFM solves the two- and three-dimensional shallow-water equations, based on the finite volume numerical method (Kernkamp et al., 2011). The 2D depth-averaged equations describes mass and momentum conservation, as presented (Deltares, 2020b):

$$\frac{\partial h}{\partial t} + \nabla.(h\boldsymbol{u}) = 0 \quad (14)$$

$$\frac{\partial h\boldsymbol{u}}{\partial t} + \nabla.(h\boldsymbol{u}\boldsymbol{u}) = -gh\nabla \zeta + \nabla.\left(vh(\nabla\boldsymbol{u} + \nabla\boldsymbol{u}^T)\right) + \frac{\tau}{\rho} \quad (15)$$

where $\nabla = \left(\frac{\partial}{\partial x}, \frac{\partial}{\partial y}\right)^T$, ζ is the water level, h the water depth, $\boldsymbol{u}$ the velocity vector, g the gravitational acceleration, v the viscosity, ρ the water mass density and τ is the bottom friction.

5.3.1.2 Model set-up

DFM allows computation on both curvilinear and unstructured grid so it is suitable for regions with complex geometry (Achete et al., 2015). In addition, it has multi-dimensional computations, especially combinations of 1D and 2D grids. This feature is efficient for considering small canals. Therefore, in this study DFM is selected for simulating floods and sediment dynamics in the Mekong Delta which comprises a dense river network and highly variable river widths, flood plains and hydraulic structures.

The large-scale hydrodynamic model of the Mekong Delta used in this study was well calibrated for the large floods in 2000 and 2001 (Thanh et al., 2020a). Unfortunately, suspended sediment data for these years were not comprehensively collected. Thus, the recent large flood in 2011 was used to validate hydrodynamics and calibrate sediment transport.

Our model setup improved prior model schematizations (Thanh et al., 2017; Van et al., 2012). The Mekong Delta is modelled using a combination of 1D networks and 2D meshes in a single computational domain. Additionally, hydrodynamics and sediment transport are computed based on an online coupling in contrast to Achete et al., (2016) who applied DelWAQ post-processing on hydrodynamic model output to calculate sediment dynamics. Compared to the model used by Thanh et al. (2020a), the present model adds sluice gates to control water flow to selected regions. These sluice gates are located along the western part of the Mekong Delta and in the Quan Lo Phung Hiep region and prevent salinity intrusion into these regions (Hoanh et al., 2009).

5.3.1.3 Grid and bathymetry

Thanh et al. (2020) describes in detail the computational grid and bathymetry presented in Figure 5.3. The grid covers the lower Mekong River from Kratie, Cambodia, to its mouths and extends to about 80 km seawards of the coastline. The dense river network of the VMD is fully represented. The floodplains in the Mekong Delta incorporated in the model are based on the flood inundation maps (Dartmouth Flood Observatory, 2004).

The computational domain consists of a multi-dimensional grid which includes 1D and 2D connections. In the Mekong Delta primary and secondary canals are represented in 1D networks while 2D cells are used for the Mekong River main channels, the floodplains and the continental shelf. The 1D network has a uniform segment length of 0.4 km while the 2D cells have a different resolution depending on the spatial scale of the locally dominant morpho- and hydrodynamic processes. Specifically, the 2D cell sizes for the Mekong River mainstreams are approximately 0.7 km in general and decrease to about 0.2 km at river bifurcations and confluences. The 2D cells are coarser for floodplains and sea areas, increasing up to around 2 km in size. The grid totally contains 73,504 cells.

Detailed bathymetries of the Mekong Delta are sparse and limited. Therefore, the bathymetry is composed based on different sources (Figure 5.3). The bathymetry of the Mekong River main channels was interpolated from cross-sectional data of the 1D-ISIS hydrodynamic model for the Mekong Delta (Van et al., 2012). The 1D network of primary and secondary canals are defined by cross-sections originally extracted from the 1D-ISIS model. For the sea areas, it is imposed from ETOPO of about 1 km resolution. The floodplain topography is obtained from the digital elevation model, with a resolution of 250 m provided by the Mekong River Commission. The estuarine branches were updated by recent in-situ data (Thanh et al., 2017).

110

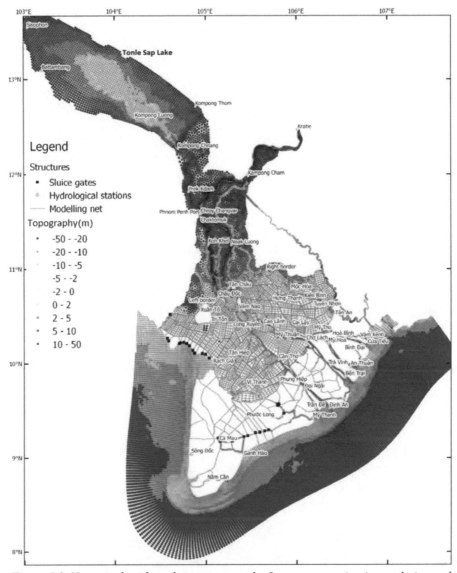

Figure 5.3. Numerical grids and river topography from cross-section interpolation and shelf topography of the Mekong Delta.

5.3.1.4 Sediment transport equation

Suspended sediment occupies the majority proportion of total sediment load in the Mekong Delta and sediment bedload is about 3% of suspended sediment load (Koehnken,

111

2014). Therefore, this research only consider suspended sediment load for modeling. Hung et al. (2014b) found that medium grain sizes of suspended sediment in floodplains fluctuate in ranges of 10 and 15 μm. In the Mekong Delta, Koehnken (2014) found a predominance of silt and clay at Kratie and Tan Chau stations, respectively. Consequently, cohesive sediment is the only sediment fraction used in this study (Thanh et al., 2017). We neglect vertical stratification, but the effect of flocculation due to salinity is included, by applying a larger fall velocity in saline water.

Suspended sediment transport is computed by online coupling between the flow and sediment transport modules of the DFM suite. Sediment transport is formulated by the 2D advection-diffusion equation for suspended sediment (Deltares, 2020a).

$$\frac{\partial c^{(l)}}{\partial t} + \frac{\partial uc^{(l)}}{\partial x} + \frac{\partial vc^{(l)}}{\partial y} - \frac{\partial}{\partial x}\left(D_x \frac{\partial c^{(l)}}{\partial x}\right) - \frac{\partial}{\partial y}\left(D_y \frac{\partial c^{(l)}}{\partial y}\right) = 0 \ (16)$$

where $c^{(l)}$ is mass concentration of sediment fraction (l) (g/l); u and v are flow velocity components (m/s); D_x and D_y are the diffusion coefficients in x and y directions respectively (m²/s).

Erosion and sedimentation in a cell are described by the well-known Krone-Partheniades equations (Partheniades, 1965):

$$E = M \left(\frac{\tau_b}{\tau_e} - 1\right) (17)$$

$$D = w_s \ c \ \left(1 - \frac{\tau_b}{\tau_d}\right) (18)$$

where E is the erosion flux (kg/m²/s), M is the erosion parameter (kg/m²/s), τ_b is the bed shear stress (N/m²), τ_e is the critical shear stress for erosion, D is the deposition flux (kg/m²/s), w_s is the settling velocity (m/s), c is near–bed suspended sediment concentration (kg/m³) and τ_d is the critical shear stress for deposition (N/m²). The equation 5 is approximated as (D = w_s c) when τ_d is much larger than τ_b (Achete et al., 2015; Deltares, 2020a; Winterwerp et al., 2006).

5.3.1.5 Boundary conditions

For hydrodynamic forcing, we defined water discharges at Kratie (the upper boundary) and water levels at the ocean (the lower boundary). In addition, the lateral offshore boundary is specified as a Neumann boundary which allows free development of cross-shore water level slopes (Roelvink and Walstra, 2004). The water discharges at Kratie are generated by measured water levels and the updated rating curve created by the Mekong River Commission (MRC). The measured water levels are collected from the near real-time hydro-meteorological monitoring system of MRC (https://monitoring.mrcmekong.org/station/014901). The water levels at the ocean are imposed by the eight main astronomical tidal constituents derived from the global tidal model of TPXO 8.2 (Egbert and Erofeeva, 2002). Local precipitation is neglected because this study focuses on the river flows.

For the lower boundary, SSC is set to 0 g/l because the river plumes are well contained within the computational grid (Thanh et al., 2017) in contrast to Manh et al. (2014) who defined the downstream boundary conditions at the Mekong River mouths from water turbidity derived from satellite images. These data have a temporal interval of around a week. However, measured data of suspended-sediment concentration at the Mekong River mouths are highly various based on tidal fluctuation. For example, (Nowacki et al., 2015) found that SSC on ebbs is considerably greater than on floods, suggesting that he boundary condition of SSC at these mouths needs a higher temporal resolution. Unfortunately, measurements at these stations with the requiered temporal frequency are not available. Therefore, the model grid was extended to completely contain the sediment plumes.

The upper boundary SSC at Kratie is not measured frequently. Therefore, we derived SSC from a regression curve with water discharges. This method is commonly used to generate SSC data in the Mekong River (Darby et al., 2016; Koehnken, 2012; Kummu and Varis, 2007; Lu et al., 2014; Lu and Siew, 2006; Manh et al., 2014; Walling, 2008). Table 5.1 shows empirically derived relationships between measured SSC and discharge for Kratie by Darby et al. (2016), Koehnken (2012) and Manh et al. (2014). Applied to measured flow at Kratie, Figure 5.4 shows the derived 2011 sediment loads. Darby et al. (2016)'s curve predicts higher SSC since it is based on a much longer period of data analysis that includes the effect of sediment load decline due to upstream dam construction (Darby et al., 2016; Kummu and Varis, 2007; Lu and Siew, 2006). Consequently, SSC generated by the Koehnken (2012) is more reliable for recent years and is used as boundary conditions.

Table 5.1. SSC rating curves for Kratie station, with SSC is suspended-sediment concentration and Q is flow discharge at Kratie station ($m^3 s^{-1}$).

Studies	SSC (mg/l)	Estimated annual sediment load in 2011 (Mt)	Analysis period
Koehnken (2012)	$0.13332 * Q^{0.7098}$	98	2011
Manh et al. (2014)	$10^{(-494.02*\log(Q)^{-4.52}+2.88)}$	96	2010-2011
Darby et al. (2016)	$0.3002 * Q^{0.8967}$	156	1981-2005

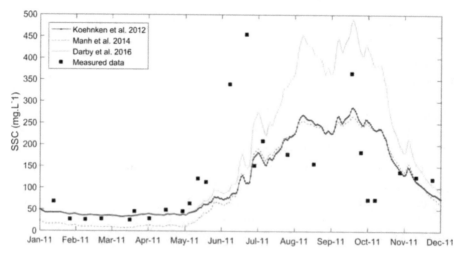

Figure 5.4. Estimated and measured suspended-sediment concentration at Kratie in 2011.

5.3.1.6 Wave modeling

Our modeling domain includes the shelf of the Mekong Delta where waves strongly influence hydrodynamics and sedimentation processes (Thanh et al., 2017). The waves at the shelf of the Mekong Delta are generated by monsoon winds (north-eastern and south-western monsoons).

Waves are computed by the Delft3D-Wave, which is a third generation SWAN model. Tu et al. (2019) calibrated this model against measured data for this region. The wave model couples with the flow model at one-hour intervals. The wave data which were derived from ERA Interim reanalysis data (https://apps.ecmwf.int/datasets/data/interim-full-daily), were imposed at the offshore boundary. The boundary conditions consist of wave height, wave period and wave direction. The wave heights off the west and east coasts of the Mekong Delta are significantly different (ADB, 2013), necessitating spatial variation in the imposed boundary conditions.

5.3.1.7 Initial conditions

Hydrodynamics in the Mekong Delta are strongly driven by the annual floods. Moreover, the Tonle Sap Lake plays a crucial role in regulating river flows the delta downstream. Water levels of the Tonle Sap Lake vary seasonally to a large extent. Therefore, correct specification of initial conditions reduces model spin-up periods. We assume that a previous flood filled the Tonle Sap Lake. Therefore, the model was spun up over the flood of 2010 and we used water levels at the end of 2010 as the initial conditions for the year 2011 simulations. Over the model domain, a uniform value of 0 g/l was set as initial conditions of SSC, since SSC is low in the low flow seasons. We used model settings for hydrodynamic parameters following Thanh et al. 2020a and 2017) including the calibrated values of the Manning roughness coefficient spatially varying in the range of 0.016-0.032. The initial bed sediment layer thickness was uniformly set at 10 m, which allows abundant sediment availability for the simulated period.

5.3.2 Sediment properties

For modeling cohesive sediment dynamics, we need to specify, critical bed shear stress for erosion (τ_{ce}), erosion rate (M) and settling velocity of sediment (w). McLachlan et al. (2017) measured shear stresses in-situ at the Song Hau estuarine branch and estimated the highest shear stress at approximately 10 Pa. However, Vinh et al. (2016) set τ_{ce} of 0.2 N/m^2 for the coastal VMD while the amplitude of τ_{ce} on the VMD floodplains fluctuated in the range of 0.028-0.044 N/m^2 (Hung et al., 2014b). M was in the range of 5.1×10^{-6} - 8.8×10^{-5} kg/m^2/s and a reasonable value for modeling is 2×10^{-5} kg/m^2/s (Hung et al., 2014b; Thanh et al., 2017; Vinh et al., 2016).

The settling velocity in the Mekong River is highly variable depending on the local hydrodynamics and salinity. Manh et al. (2014) mention that the calibrated w value in the main channels of the Mekong River was 1.3×10^{-3} m/s. Hung et al. (2014) calculated that settling velocities on the VMD floodplains fluctuated from 2.2×10^{-4} to 1.8×10^{-3} m/s. McLachlan et al. (2017) estimated settling velocities on the Song Hau estuarine branch to be much smaller, with an average magnitude of around 5×10^{-5} m/s. Furthermore, w is also influenced in saline waters by growing flocs. For instance, Wolanski et al. (1996) observed that the median size of flocs in the Mekong estuary ranges between 50 to 200 μm and this changes the sediment settling velocity. Vinh et al. (2016) revealed that w in fresh and saline waters were 5×10^{-5} and 3.25×10^{-4} m/s, respectively.

5.3.3 Model performance evaluation

Percent bias (PBIAS) and index of agreement (Skill) are commonly used statistical indices to evaluate model performance (Achete et al., 2015; Ferré et al., 2010; Ji, 2017; Van Liew et al., 2007; Thanh et al., 2017). These indices are calculated as

$$PBIAS = \frac{\overline{S-M}}{\overline{M}} \ (19)$$

$$Skill = 1 - \frac{\Sigma(S-M)^2}{\Sigma(|M-\bar{O}|+|O-\bar{O}|)^2} \ (20)$$

where S and M are simulated and measured SSC, respectively; and $\bar{M}$ and $\bar{O}$ are time average measured SSC.

PBIAS and Skill were used to assess simulated discharge and SSC at mainstream stations. PBIAS values present the average tendency of simulated results. A perfect PBIAS value of 0 illustrates that modelled results are completely unbiased. Positive and negative PBIAS values indicate model biases toward overestimation and underestimation, respectively. Skill was introduced by Willmott (1981) and it presents how accurate the model estimates the variation in measured data. Skill values range from 0 to 1 in which the value of 1 indicates that simulations and observations have perfect agreement while the value of 0 shows disagreement between them. A well calibrated model should have values of $|PBIAS| < 0.25$ and $Skill > 0.2$ (Ji, 2017; Moriasi et al., 2007). Table 5.2 depicts categories of model performance intervals.

Table 5.2. Qualification of model performance indicated by PBIAS and Skill indexes.

Qualification	$|PBIAS|$	Skill
Excellent	< 0.1	$1.0 - 0.65$
Good	$0.1 - 0.15$	$0.65 - 0.5$
Reasonable/fair	$0.15 - 0.25$	$0.5 - 0.2$
Poor	> 0.25	< 0.2

5.4 RESULTS AND DISCUSSION

5.4.1 Model calibration and validation

5.4.1.1 Hydrodynamic and salinity calibration

Detailed results of model performance for water discharge at stations on the mainstream of the Mekong River is illustrated in Table 5.3. These stations are selected to validate the model since the data are available for the chosen year. However, these stations can represent flood propagation along the Mekong River as they are located from the upstream boundary (Kratie) to the strongly tide-dominated areas (Can Tho and My Thuan). The model reasonably simulates water discharge in the delta because the values of statistical indexes at these stations are higher than the reasonable value, except Chau Doc station. Skill values of these stations are higher than 0.8 which classifies as excellent.

Generally, the model slightly underestimates water discharge as PBIAS values are negative.

Table 5.3. Statistical indexes of model performance of water discharge, suspended-sediment concentration and sediment flux.

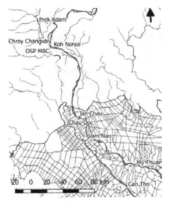

Station	Discharge		SSC		Daily Sediment Flux	
	PBIAS (%)	Skill	PBIAS (%)	Skill	PBIAS (%)	Skill
Chroy Changvar	-6	0.98	25	0.63	11	0.83
Koh Norea	-16	0.86	N/A	N/A	-7	0.86
OSP MRC	N/A	N/A	5	0.72	-8	0.91
Prek Kdam	N/A	N/A	N/A	N/A	-40	0.56
Tan Chau	-4	0.99	61	0.90	36	0.93
Chau Doc	-33	0.85	-18	0.78	-45	0.67
Vam Nao	N/A	N/A	-2	0.90	N/A	N/A
My Thuan	-18	0.94	16	0.94	-23	0.86
Can Tho	-12	0.97	7	0.87	-20	0.73

The fall velocity of cohesive sediment is influenced by salinity, which enhances flocculation processes (Mhashhash et al., 2018; Portela et al., 2013). In the Mekong Delta, Wolanski et al. (1996) found that sizes of flocs in the saltwater region are much larger than those of suspended sediment grains. The length of saltwater intrusion into the Mekong River is approximately 50 km from the river mouth (Nguyen and Savenije, 2006; Nowacki et al., 2015; Wolanski et al., 1998). Salinity intrusion into the Song Hau is limited by seasonally varying river flow (see An Lac Tay station in Figure 5.5). The largest salinity intrusion occurs during the low flow season, while the water at the river mouths is nearly fresh in the high flow seasons (Wolanski et al., 1996). As a result, the salinity is only measured in the low flow seasons, so the calibration period of salinity did not include the period Jul-Dec 2011.

Figure 5.5 shows that simulated and measured salinity are in reasonable agreement. The 2D model is capable of modeling the seasonal and tidal cycle variations of salinity. Specifically, the highest salinity at Tran De station is about 20 ppt during the low flow season. We calibrated salinity intrusion by adapting the horizontal eddy diffusion coefficient leading to a value of 450 m^2s^{-1} constant over the model domain. This high value was also found in other modeling studies and is caused by considerable sub-grid-scale processes (Talley et al., 2011).

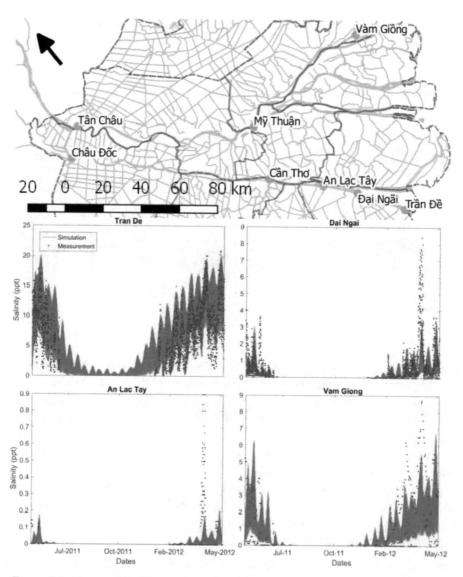

Figure 5.5. Measured (in blue) and simulated (in red) salinity at Tran De, Dai Ngai, An Lac Tay and Vam Giong.

5.4.1.2 Sediment dynamics calibration

To calibrate the sediment dynamics model, we adapted and modified the proposed approach of alternative settings developed by Van Maren and Cronin (2016). Specifically, the settling velocity and the critical shear stresses were estimated based on available measured data. The settling velocities in fresh water and saltwater are well measured and applied in numerical modeling (Le et al., 2018; Thanh et al., 2017; Vinh et al., 2016). We calibrated to lower the sediment flux by increasing the critical shear stress and decreasing the erosion rate. The model performances in simulating SSC and sediment flux are presented in Figure 5.6, Figure 5.7 and Figure 5.8.

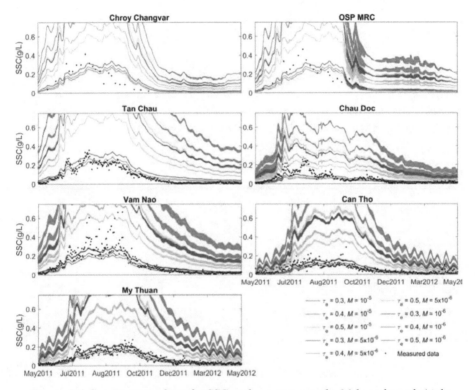

Figure 5.6. Sensitivity analysis for SSC at the stations on the Mekong branch (right panels) and the Bassac branch (left panels).

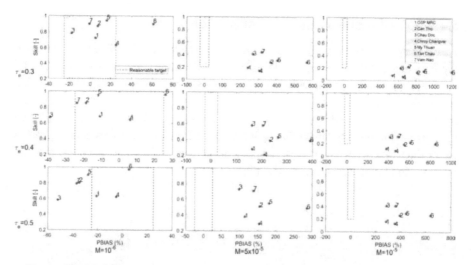

Figure 5.7. Model performance of SSC in the sensitivity analysis. These simulations were set up with the same settling velocity, changing τ_{ce} and M parameters in the range presented in rows and columns, respectively. The target square presents the acceptable level of model performance. The locations of these stations are indicated in Figure 5.1.

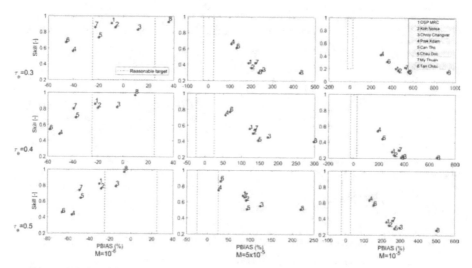

Figure 5.8. Model performance of sediment fluxes in the sensitivity analysis. These simulations were set up with the same settling velocity, changing τ_{ce} and M parameters in the range presented in rows and columns, respectively. The target square presents the acceptable level of model performance. The locations of these stations are indicated in Figure 5.1.

The model was calibrated against measured SSC data in the high-flow 2011 season and the low-flow 2012 season focusing on stations along the Mekong River. The order of the low-flow and high-flow seasons is of great importance. This order not only plays a considerable role in controlling seasonal variations of the Mekong River flow, but also in sediment trapping. We select the low-flow season after the high-flow season because of the regulation of the Tonle Sap Lake.

The parameters of roughness, τ_{ce}, M and w have considerable impacts on sediment dynamics (Achete et al., 2015; Manh et al., 2014). Roughness coefficients are not an efficient calibrated parameter for sediment model because they are evaluated in hydrodynamic calibration and this eliminates a free variable in sediment calibration (Gibson et al., 2017). The settling velocity was not considered as calibration parameter, because it is well measured and successfully used in other numerical studies (Gratiot et al., 2017; Le et al., 2018; Marchesiello et al., 2019; McLachlan et al., 2017; Thanh et al., 2017; Tu et al., 2019; Vinh et al., 2016). The w was set to 5×10^{-5} m/s and 3.5×10^{-4} m/s for fresh and saline waters, respectively, with interpolated values for brackish environments. Recent measurements by Gratiot et al. (2017) and Le et al. (2018) have reasonable agreement with these settling velocities.

The calibration ranges of τ_{ce} and M were chosen from measurements of prior studies (e.g. Berlamont et al., 1993; Hung et al., 2014a; Manh et al., 2014; McLachlan et al., 2017; Vinh et al., 2016). The selected ranges for the two parameters were for τ_{ce} 0.3 - 0.5 N/m^2 and for M $10^{-5} - 10^{-6}$ kg/m^2/s.

In general, calibration simulations overestimate SSC on the Mekong River (Figure 5.9). The model clearly produces seasonal variations of SSC that are strongly dominated by the annual floods, which is reflected by high skill values (Table 5.3). In addition, SSC also varies with spring-neap cycles at Can Tho and My Thuan stations where tidal influence is high. Within the selected calibration range, M has a much stronger influence on SSC than τ_{ce} and this can be explained by the bed erosion flux computed by the equation 4. The curves and peaks timing of simulated SSC resemble observed SSC. It is noted that the smallest τ_{ce} and the highest M in the selected ranges result in unrealistic SSC (> 1 g/l). SSC at Chau Doc is underestimated probably due to the underestimation of water discharge at Chau Doc station (Table 5.3).

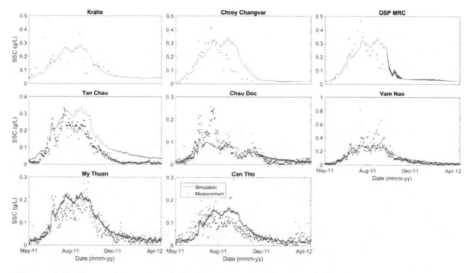

Figure 5.9. Comparison of modelled and measured suspended-sediment concentration.

The measured sediment fluxes are estimated from daily average discharge and SSC at stations on the mainstream Mekong River (Figure 5.9 and Figure 5.10) and compare well with modeled behavior (Figure 5.10). The model parameter set which results in the best fit of simulated and measured suspended-sediment concentration and sediment transport, has a τ_{ce} value of 0.3 N/m2 and a value for M of 10^{-6} kg/m^2/s.

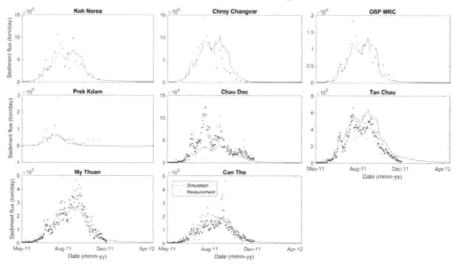

Figure 5.10. Comparison of modelled and measured suspended sediment flux.

During the calibration process of SSC, we found that the varying dominant hydrodynamic factors across the model domain, strongly influence the results of model calibration. For example, the run which has τ_{ce} of 0.6 N/m^2 and M of 8 x 10^{-5} kg/m^2/s, compares well with measured data at the fluvial-dominant stations while it highly underestimates SSC at the tide-dominant stations. The initial SSC and bed sediment availability had a very limited impact on the calibration. This is in contrast with Achete et al. (2015) who found that initializing the model with bed sediment could cause a high SSC and take around 5 years to be reworked and with van Kessel et al. (2011) who revealed that a simulation with no bed sediment could take up to 3 years in order to reach the equilibrium conditions. For our study, Figure 5.10 shows that the simulation period begins in the low flow season (from May 2011) at which the SSC is low, so the model takes a short spin-up time of around two weeks. Bed-sediment availability is essential to skillfully model SSC in the Mekong Delta. For example, at some stations (e.g. Chroy Changvar and OSP MRC) SSC is slightly higher than those at Kratie (the only source of sediment in modeling). Probably the abundance of sediment available in the Mekong River bed makes model calibration less subject to bed sediment definitions and initial SSC as reported by Achete et al. (2015) and van Kessel et al. (2011).

5.4.2 Hysteresis relations of suspended-sediment concentration and water discharge

Our model is able to reproduce SSC hysteresis during a river flood (Figure 5.11 and Figure 5.12) with SSC being higher during the rising phase of the river flood than during the falling tide of the river flood. Figure 5.11 clearly indicates the characteristic clockwise loops at different stations (Williams, 1989), with the SSC peak occurring earlier than the discharge peak. A general mechanism for SSC hysteresis is early suspension of easily erodible sediments at the start of the river flood (Landers and Sturm, 2013). Walling (2008) observed the SSC hysteresis which reflects sediment remobilization of the Mekong River. However, our modeling effort did not define that process since we applied a single sediment fraction with constant properties throughout the model runs. Also, the SSC hysteresis does not stem from the boundary since SSC at the boundary was defined by a direct relationship between river flow and SSC at Kratie (Figure 5.12).

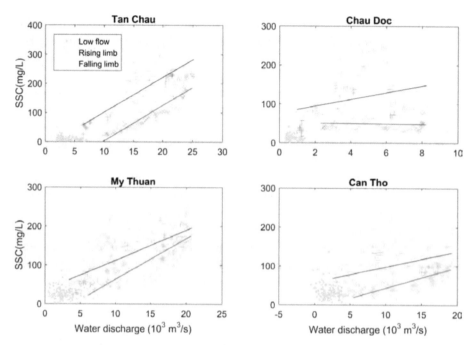

Figure 5.11. Relationship between daily averaged measured suspended-sediment concentration and water daily averaged discharge at four stations, namely Tan chau, Chau Doc, My Thuan and Can Tho. Low flow conditions are from January to June, 2011. The rising phase begins from June to the yearly discharge peak in September, whereas the falling phase is taken from September to January 2011.

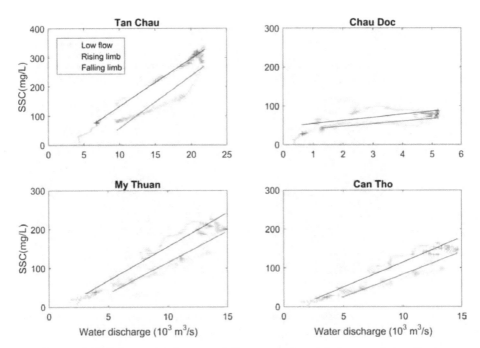

Figure 5.12. Relationship between daily averaged modeled suspended-sediment concentration and water daily averaged discharge at four stations, namely Tan chau, Chau Doc, My Thuan and Can Tho. Low flow conditions are from January to June, 2011. The rising phase begins from June to the yearly discharge peak in September, whereas the falling phase is taken from September to January, 2011.

Instead, we found that the main factor causing modelled SSC hysteresis is the sediment trapping of the Tonle Sap Lake. The sediment trapping decreases the SSC of outflows significantly compared to inflows. During a rising river flood, flood flow with high SSC from the Mekong River diverts to the Tonle Sap River at Prek Kdam to fill the Lake (Figure 5.13a). The sediment largely deposits in the lake. For example, Kummu et al. (2008) found that around 80% of sediment which is stored in the lake and its floodplains, is from the Mekong River and tributaries. In the late high-flow season, the flow of the Tonle Sap River reverses when water levels on the Tonle Sap Lake are higher than those on the Mekong River (Fujii et al., 2003; Kummu et al., 2014b; Thanh et al., 2020a). Although SSC on the Mekong River is still high (~200 mg/l), the low SSC water from the Tonle Sap River (~20 mg/l) mixes with the Mekong River flow at the Phnom Penh confluence reducing SSC in the confluence downstream (Figure 5.13b). Our model adequately captures these sediment dynamics (Figure 5.14). Our finding confirms estimates by Kummu et al. (2008)'s of an annual sediment deposition of about 5.7 Mt in the Tonle Sap Lake.

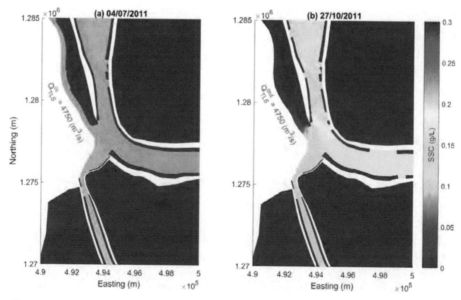

Figure 5.13. Spatial variation of modelled suspended-sediment concentration at during inflows (a) and outflows (b) of the Tonle Sap River.

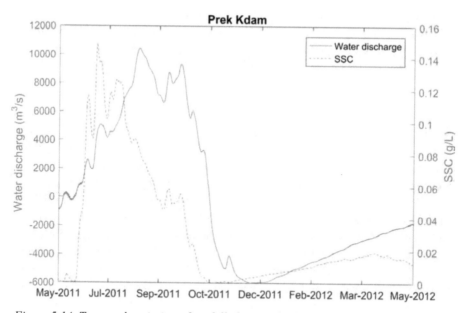

Figure 5.14. Temporal variation of modelled water discharge and suspended-sediment concentration at Prek Kdam (Tonle Sap River).

It should be noted that our model did not consider tributaries of the Tonle Sap catchment. The tributaries contribute up to about 30% of the inflow into the Tonle Sap Lake (Kummu et al., 2008) and supply a sediment amount of about 2 Mt/y (Kummu et al., 2008; Lu et al., 2014). With additional flows from Tonle Sap tributaries, outflows from the Tonle Sap Lake would increase slightly, with slightly higher SSC, increasing sediment fluxes from the Tonle Sap River to the Mekong River. The connection to the Tonle Sap Lake plays a critical role in regulating flows in the Mekong Delta and the Tonle Sap Lake received about 3.7 Mt in 2011 which resulted from differences between inflows and outflows of the Tonle Sap Lake. The discrepancies between inflows and outflows of the Tonle Sap Lake are still under discussion. Recent studies show opposite results on sediment transport of the Tonle Sap Lake (Kummu et al., 2008; Lu et al., 2014). They used measured discharge and SSC to investigate whether the Tonle Sap Lake receives sediment from or supplies sediment to the Mekong River. Interestingly, Kummu et al. (2008) found that the lake receives a net sediment amount of about 5.7 Mt/y, in which sediments are transported from the Mekong River and Tonle Sap tributaries around 7 Mt/y and supply about 1.38 Mt/y to the Mekong River in the outflow period. In contrast, Lu et al. (2014) estimated that the mean sediment inflow and outflow are 6.3 Mt and 7 Mt, respectively. This means the Tonle Sap Lake supplies about 0.7 Mt. This estimate may be incorrect due to a limitation of data used and this study is opposite to some estimates (Koehnken, 2012, 2014; Kummu et al., 2008; Manh et al., 2014).

5.4.3 Seasonal variation of suspended sediment

This section describes spatial and temporal variations of modelled SSC in the Mekong Delta which consists of different hydrodynamic regions. Figure 5.15 shows hourly simulated SSC at some selected stations in the Mekong River. In general, SSC at these stations vary significantly throughout the selected period. The SSC are highest in August and September, and lowest in March. These variations are strongly dominated by the annual floods, so they have similar seasonal variability, but the magnitudes are different between these stations. The differences result from the sediment availability in the channel system. For instance, the SSC peaks at Kampong Cham (upper station) are slightly lower than that at Chroy Changvar and Tan Chau (lower stations). This implies that there is an additional source of sediment which affects SSC in the Mekong River. The reason is that sediment modeling included sediment availability in the channel system. This is in line with the presence of bed-sediment availability within the Mekong channel system, CMD, as observed by Walling (2008).

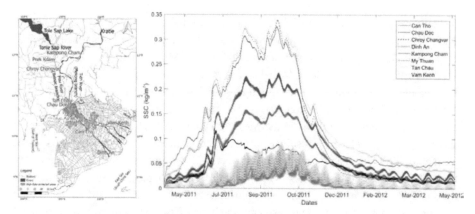

Figure 5.15. Separated regions in the Mekong Delta based on hydrodynamic conditions and SSC variation in these regions.

In the CMD, SSCs at Kampong Cham increase rapidly coinciding with flows in the high flow season. The highest value is about 0.35 g/l in the high flow season while SSCs fluctuate around 0.05 g/l in the low flow season. The river flow is the dominant hydrodynamic factor in this region, so SSCs fluctuate with the river flow. Besides, the floodplains in this region also have insignificant impact on SSC. Downstream of the Tonle Sap-Mekong confluence, SSCs are considerably influenced by interaction of the Tonle Sap River and Mekong River. On the Tonle Sap River, SSCs of the inflows (Mekong River to Tonle Sap Lake) are significantly higher than in the reversal flows, reflecting the sediment transport to and efficient trapping of the Tonle Sap Lake.

The VMD receives sediment from the Song Tien, the Song Hau and Cambodian floodplains. At Tan Chau, SSC variations are high in the high flow season, comparable to SSCs at Chroy Changvar while they show tidal variations during the low flow season. In the VMD middle, the tides strongly drive sediment fluctuations in both the high flow and low flow seasons. In the high flow season, the highest SSC can reach 0.25 g/l at My Thuan and 0.2 g/l at Can Tho. In the low flow seasons, SSCs obviously vary with spring-neap cycles, fluctuating from 0.025 to 0.05 g/l. These values are completely consistent with the analysis by Dang et al., (2018). SSC during ebb tides are marginally higher than those during flood tides. This asymmetry causes seaward flux of suspended sediment. At the mouths the Mekong River, SSCs fluctuate in range of 0.01 - 0.1 g/l in the high flow season, coinciding with tidal variations while they are smaller than 0.05 g/l in the low flow season. These simulated values of SSC are slightly lower than measured data analyzed by McLachlan et al., (2017). There are several factors contributing to the differences such as salinity stratigraphy, estuarine turbidity maximum and flocculation. The salinity stratigraphy was neglected since a 2D depth-averaged model setup was used while the flocculation was taken into modeling by changing settling velocities in freshwater and saltwater.

5.4.4 Sediment budget

The Mekong River at Kratie supplied more than 99 Mt of suspended sediment that was transported towards the Mekong Delta from June 2011 to June 2012. Based on our model validation at specific sites, we can now derive a sediment budget describing the distribution of these sediments within the Mekong Delta, illustrated in Figure 5.16.

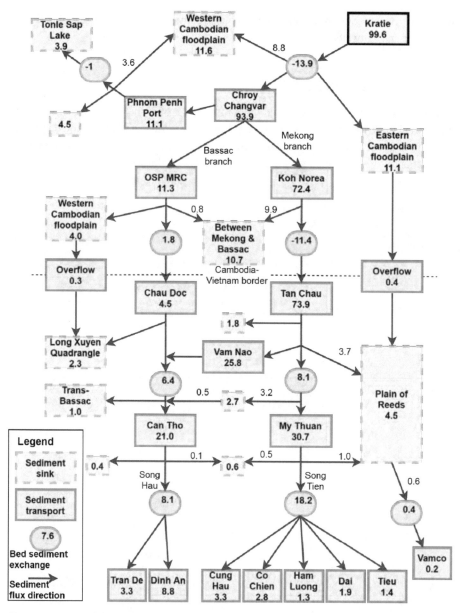

Figure 5.16. Modelled sediment budget (in Mt) of the Mekong Delta, location names are indicated in Figure 5.1 and Figure 5.3.

The river flood erodes the river channel of the Mekong from Kratie to Phnom Penh by around 13.9 Mt of sediment while the adjacent floodplains receive an amount of 8 Mt

130

(northern floodplain) and 11.1 Mt (southern floodplain) during the high-flow season. Approximately 94 Mt flows into the Mekong Delta at Phnom Penh. At Phnom Penh, the Mekong River connects to the Tonle Sap River which transports 11.1 Mt. During rising river flood, the Tonle Sap River transports 4.5 Mt to the Lake, whereas the Tonle Sap River transports 0.6 Mt to the Mekong River during the falling river flood. The net result is a supply of 3.9 Mt to the Lake. Downstream of Phnom Penh the Mekong River separates into two branches, namely Mekong and Bassac (see Figure 5.1), transporting 72.4 Mt (73 %) and 11.3 Mt (11 %) seaward, respectively. An additional amount of 0.7 Mt (0.7 %) is delivered over the floodplains between Vietnam and Cambodia. This portion is slightly higher than estimates (64-71%) of Manh et al. (2014). The difference is probably due to the inclusion of bed-sediment availability in this study.

The percentage of the total sediment supply at Kratie transported into the VMD by the main channels depends on water years. In wet years, the percentage is smaller than this in the dry years. The sediment trapping of the Cambodian floodplain and Tonle Sap system in the wet years is higher than that in the dry years (Manh et al., 2014).

Suspended sediments are transported into the VMD by the Mekong branch (Tien River), the Bassac branch (Hau River) and floodplain flows. Among of these ways, Tien River is the major way which conveys about 74 Mt at Tan Chau, accounting for 93% of the total sediment discharge towards the VMD. The overland flows transport small amounts of sediment during the high-flow seasons.

In the VMD, river flows and sediment transport are strongly affected by the dense man-made canal system so sediment transport in the VMD is complicated, especially the interaction between the Tien River, the Hau River and the floodplains. The canals between the Tien River and Hau River divert water and sediment from the Tien River to the Hau River due to slightly higher water level in the Tien River. The water discharge in the Tien River and Hau River becomes similar from the connecting canal (Vam Nao) seaward (Dang et al., 2018b; Thanh et al., 2020a). The Vam Nao canal diverts ~25.8 Mt from the Tien River to the Hau River. However, sediment fluxes at My Thuan (30.7 Mt) are slightly higher than at Can Tho (21.0 Mt).

The differences in sediment fluxes throughout these two stations results from slightly higher SSC in the Tien River compared to the Hau River (presented in Figure 5.9). The ratio of sediment fluxes between the both stations is equivalent to recent studies (e.g. Dang et al., 2018; Manh et al., 2014), but magnitudes of sediment fluxes are somewhat lower. The lower sediment fluxes may come from different period in estimation as we considered sediment fluxes in the low-flow season 2012 while Dang et al. and Manh et al. considered the low-flow season 2011. Dang et al. (2018) estimated that the sediment fluxes in 2011 at My Thuan (38.3 Mt) and at Can Tho (23.4 Mt). These values are reliable because they are derived from daily measured data.

At the river mouths, the Mekong River delivers an amount of 22.8 Mt to the sea in which it transports 10.7 Mt and 12.1 Mt, respectively, through the Tien and Hau branches. The Hau branch exports about 8.8 Mt and 3.3 Mt via the Dinh An and Tran De mouths respectively. Besides, the Tien River transports approximately 10.7 Mt of sediment by the Cung Hau (3.3 Mt), Co Chien (2.8 Mt), Ham Luong (1.3 Mt), Dai (1.9 Mt) and Tieu (1.4 Mt) mouths. Although water volume discharged by the Tien River's mouths is slightly higher than that by the Hau River's mouths (Thanh et al., 2020a), suspended sediment exported though the Tien River is smaller than the Hau River. This can be explained by the Tien River has a larger depositional plains compared to the Hau River. This is determined by sediment deposition of about 18 Mt and 8 Mt from My Thuan (the Tien River) and Can Tho (the Hau River) to the coast, respectively. Previous studies did not directly compute sediment fluxes at the river mouths of the Mekong River, but computed transports at other stations on the mainstreams or extrapolate measured data (Dang et al., 2018b; Nowacki et al., 2015). It was assumed that all sediment of the Mekong River would be transported to the sea. This study included riverbed deposition and erosion processes throughout the entire Mekong Delta for the first time.

Riverbed sediment exchange is difficult to measure especially throughout such a large domain. Our model suggests that riverbed erosion occurs in Cambodia while deposition happens in Vietnam. Unfortunately, observed data of riverbed change are unavailable to validate these results. They are also strongly affected by human activities, such as sand mining. Sand mining is probably much higher than the sedimentation rate of the Mekong River in Vietnam (Brunier et al., 2014). Moreover, estimated sand mining in the Hau River is about 7.75 million m^3 in 2011 (Bravard et al., 2013) (~ 12.4 Mt, based on a bulk density of 1600 kg/m^3) which is about one-third of the sedimentation rate of about 39.6 Mt. The sediment dynamics at the river mouths have seasonal variations. During the high flow season, the Mekong River supplies a substantial amount of sediment to the sea due to seaward residual velocity. During the low flow season, the tidal processes cause a small amount of landward sediment import (Gugliotta et al., 2017; Nowacki et al., 2015; Xing et al., 2017). The landward residual sediment flux is resulted by baroclinic effects (Nowacki et al., 2015). The modelled sediment fluxes at the river mouth stations capture characteristics of sediment exchange due to tidal processes.

5.5 CONCLUSIONS

Recent measurements of suspended sediment transport into the Mekong Delta show that the annual sediment transport is significantly lower than the commonly used value of 160 Mt. The suspended sediment loads are highly varied due to variations of water discharges.

We used a 1D-2D numerical model to simulate the hydrodynamics and suspended sediment transport throughout the Mekong Delta, forced by rivers, tides and waves. The modeling grid covers the entire Mekong Delta, the connected Tonle Sap Lake and its shelf. This grid considers the interaction between the Mekong River and the sea. In addition, the grid includes a dense network of rivers and man-made canals, considering hydrodynamics and sedimentation in the floodplain regions. Our modeling effort reproduces the hydrodynamics and SSC and transport. In sediment modeling, the erosion rate is an important parameter when considering bed-sediment availability.

This is one of the first studies which are able to investigate sediment dynamics at a large scale of the entire Mekong Delta. The model reproduces the suspended sediment concentrations and sediment fluxes at several stations located in different hydrological regions. Thus the model can simulate sediment dynamics in the Mekong Delta accurately. The presence of bed-sediment availability, as a source of sediment supply, allows sediment exchange between river bed and water column. Additionally, bed-sediment availability is a cause of sediment hysteresis. Noticeably, we found that another cause of hysteresis effects is efficient trapping of the Tonle Sap Lake. The inflow of the Tonle Sap Lake has higher suspended sediment concentration compared to the outflow, causing clockwise loops of the hysteresis effects.

Our study confirms that the annual sediment load of the Mekong River into its delta is much lower than the common estimate. The annual sediment load in 2011-2012 at Krate, Cambodia is approximately 99.6 Mt. About 79% of the annual sediment load is transported to the VMD, but only 23% of the total sediment is exported to the East Sea. This determines that trapping efficiency of the VMD system is generally high.

This study indicates that numerical models are a useful and efficient tools to gain a better understanding of hydrodynamics and sediment transport in large-scale areas. They are more helpful in the cases of high variability of channel widths and large spatial scales. Due to taking the dense network of rivers and canals onboard, the model is likely impossible to apply at long time scales, such as centuries. Understanding of sediment sources, pathways and transport dynamics not only encourages to investigate morphodynamics, but also helps to address challenges of negative impacts on ecology and develop sustainable management strategies. Particularly, the results can be necessary and useful inputs for the Mekong Delta Plan.

6

CONCLUSIONS AND RECOMMENDATIONS

This chapter shows conclusions and recommendations. The concluding section presents findings of this research and answers of the research questions. Recommendations for further studies are also discussed.

6.1 CONCLUSIONS

6.1.1 Introduction

This study aims to investigate sediment dynamics and sediment budget in the Mekong Delta by using a process-based model. Reproducing sediment dynamics in the Mekong Delta is a challenge due to limited data availability and its complex system. Previous chapters described an efficient interpolation method for sparse and limited data and the different modeling efforts in detail leading to a skilful reproduction of observed sediment dynamics in the Mekong Delta. This section provides answers to the research questions mentioned in the Introduction chapter.

6.1.2 Answer to research questions

  1. How can a 2D/3D model be applied in cases of limited topography data?

A 2D/3D model requires channel topography with high spatial resolution to accurately represent the channel-bed surfaces. In cases of limited topography data, it needs an interpolation method to increase spatial resolutions. The usual interpolation methods are applied for isotropic data. However, channel topographies are often strongly influenced by the flow direction and the presence of the thalweg. Consequently, applying usual interpolation methods on the channel topography can cause unrealistic bed surfaces (e.g. discontinuous thalweg and jags) which gives rise to errors in interpolated topographies in limited data cases and increased uncertainty in modeling outputs, especially for sediment dynamics and morphological modeling. Therefore, limited data of topographies can only be used for a 2D/3D model if their spatial resolution is artificially increased by an advanced interpolation method. This study proposes an interpolating approach for generating channel topographies. The approach consists of three steps: (1) Anisotropic bed topography data locations are transformed to an orthogonal and smooth grid coordinate system that is aligned with the river banks and thalweg. (2) Sample data are linearly interpolated to generate river bathymetry. (3) The generated river bathymetry is converted into its original coordinates. This approach reveals that elimination of anisotropic effects of the channel topographies is a critical step and it suggests to use the linear interpolation method.

  2. What is the role of coastal processes in sediment modeling for the Mekong Delta?

The Mekong Delta is strongly influenced by coastal processes such as tides and waves. The Mekong River delivers sediment to the sea where the depositing sediment forms the Mekong Delta. Besides, salinity also affects sediment dynamics since it can change settling velocity of sediment particles in the water column while spatial salinity gradients generate tide residual currents. Changes of suspended sediment concentration in the Mekong coastal area are not only influenced by the tide but also highly dependent on the river flows. The seasonal patterns of sediment plume dynamics show north-eastward flow in August and September while the plume is more variable in October and clearly shifts to south-westward flow in the low flow season. Waves play an important role in sediment stirring. The sediment deposited during the high flow season, mostly on the edge of the delta front, is strongly resuspended by waves and currents in December and January. Reduction of riverine sediment supply in the low flow season leads to strongly decreased concentrations throughout the delta. Meanwhile, nearshore SSC remains relatively high due to the combined effect of wave stirring and the cross-shore gravitational circulation with an onshore-directed near-bed current. Excluding salinity gradients spreads the sediment over the shelf, because in that case there is no reduction in the vertical diffusivity due to buoyancy effects and no gravitational circulation.

3. *How does the delta-based water infrastructure influence hydrodynamics in the Mekong Delta?*

Recently, the Mekong Delta has been considerably modified by human interventions, particularly water infrastructures. Over the last two decades high dykes were built in the VMD after the high damage flood of 2000. Generally, the high dykes were designed at a crest level of 0.5 meter above the year 2000 flood peak. As a result, the hydrological regime in the VMD has been influenced by the high dykes because they prevent flood water flowing into the floodplain area protected by these high dykes. This study evaluated the effects of high dyke construction on downstream hydrodynamics by using a numerical model for the year 2000. A coherent set of dyke development scenarios for the VMD was defined in order to quantify possible effects. The scenarios consider high dyke development in the VMD floodplain regions, including the LXQ, PoR and Trans-Bassac regions.

The dyked floodplains in the LXQ, PoR and TransBassac regions cause increases of 12.3, 6.1 and 1.1 cm of annual mean water levels at Chau Doc station, respectively. With the recent development of high dyke construction until 2011, the yearly averaged water levels would increase by 10.2 cm (Chau Doc), 1.5 cm (Long Xuyen) and 0.2 cm (Can Tho) on the Song Hau. The annual mean water levels would rise up to 22.3 cm at Chau Doc station if the entire VMD is fully protected by high dykes, while water levels on the Song Tien are less affected by the high dykes. Among the mentioned regions, the PoR has the highest effect on Song Tien's water levels since they are directly linked. Although the LXQ is not directly linked to the Song Tien, it causes rising water levels by around 3.6 and 1.1cm at Tan Chau and My Thuan, respectively. The mean water level at Tan Chau could increase by 16.9 cm if the VMD's floodplains are fully dyked. Besides the water levels the high dykes also influence the tidal propagation along the Mekong branches due to

137

reduction of water retention during the high flow season. The high dykes avoid floodwaters to come into the protected floodplains, causing an increase and a decrease of water discharge in the high flow and low flow seasons respectively. This change of hydrograph influences the tidal propagation in the VMD.

4. What are prevailing sediment dynamics and sediment budget in the Mekong Delta?

Investigations of sediment dynamics for the Mekong Delta were a challenge due to limited availability of sediment data. Sediment dynamics and sediment budgets are usually derived from in-situ measurements. This needs a number of measured stations and campaigns which are not available in the Mekong Delta. In that case, numerical models are useful tools to derive sediment dynamics and sediment budgets. A 1D-2D numerical model was developed to simulate the hydrodynamics and suspended sediment transport throughout the Mekong Delta, forced by river discharge, tides and waves. The set-up for sediment modeling was improved in order to consider important features, such as sediment plumes and at-bed sediment exchange. After calibration, the model is capable of capturing hysteresis relationships between water discharge and SSC. The hysteresis relationships are caused by flow regulation of the Tonle Sap Lake in which inflow water contains suspended sediment much lower than outflow water, leading to a trapping amount of 3.7 Mt in 2011. Analysis of the modelled sediment budget reveals that sediment transport into the Mekong Delta even in a high water year is much lower the common estimate of 160 Mt (Chapter 5). The Mekong River supplies ~99 Mt of suspended sediment to the VMD from June 2011 to June 2012 and only 23% of this amount is discharged into the sea. The remaining part of the sediment is trapped in the Mekong Delta. An amount of ~79.2 Mt/year is delivered to the VMD by the Song Tien, the Song Hau and overflows. Analysis of modelled sediment dynamics over the entire Mekong Delta, considers riverbed sediment exchange. The modelled results revealed that the Mekong mainstream riverbed is eroded in Cambodia while deposited in Vietnam. The sediment deposited in the channel system in the VMD is significantly higher than that in the floodplains.

6.2 RECOMMENDATIONS

The previous chapters describe the setup, validation and analysis of an unstructured 2DH process-based model covering the entire Mekong Delta area between Kratie and the East Sea, with the aim of investigating sediment dynamics and sediment budget in the entire Mekong Delta. However, the sediment model was calibrated for the water year 2011 so it may not reach the equilibrium conditions. To reach that state it may need to simulate a period of 3 years (e.g. van Kessel et al., 2011). The channel topography used in modeling was reproduced from channel cross-sections by an interpolation method (Thanh et al.,

2020b) so elevations of the river bank may not be generated accurately. The elevations significantly affect rivers overflowing their banks during the high flow season. In addition, the model was calibrated well to SSC and suspended sediment flux at the stations located in the VMD upstream. The model possibly needs to be calibrated at onshore/offshore stations. The modelled results suggest several aspects that are considered important for further research.

Coupling with 3D modeling

The unstructured 2DH process-based model for the entire Mekong Delta was used to analyse sediment transport dynamics. The VMD experiences salinity intrusion and effects of salinity on sediment settling velocity were considered. However, the Mekong estuaries experience density driven flows by spatial gradients in salinity leading to tide residual circulation patterns. Without effects of salinity, the sediment will be spread farther to the sea (Thanh et al., 2017). In addition, no salinity changes characteristics of the VMD estuarine turbidity maxima (ETM), leading to change sediment trapping in the ETM. The ETM forms a sediment cloud in the water column which can be the result of sediment resuspension processes (Wu et al., 2012). This requires to incorporate 3D modeling for further investigation.

Interaction with mangrove forests

The presence of mangroves in the Mekong estuarine system is important for a number of reasons. Tidal pumping by flow asymmetry in the mangrove fringes of the estuary coastlines and river banks provides a mechanism for sediment import into these forests (e.g. Furukawa et al., 1997; Furukawa and Wolanski, 1996; Mullarney et al., 2017). Moreover, the retention of water in the forest during flood impacts on tidal asymmetry and increases the flow velocity during the ebbing tide in the adjacent channels, providing a mechanisms for bank erosion and bed scour (e.g. Mazda et al., 1995; Wu et al., 2001). Thirdly, mangrove shrubs and tree trunks provide an effective means to dissipate incident wave energy along the coast (e.g. Mazda et al., 2006; Quartel et al., 2007), locally reducing the potential for sediment resuspension and transport. All these phenomena have profound effects on the amount of sediment that is available for resuspension and transport along the coastal fringes in our modeling domain. As such, vegetation feedback effects on currents and waves should be incorporated in the next modeling cycle.

139

Morphodynamics

This thesis focused on simulating the spatio-temporal distribution of suspended cohesive sediments. The logical next step is to look into the mechanisms for bedload transport and ultimately, to compute bed level updates and the feedback of morphological changes on the hydrodynamics.

Delta management and planning

In this thesis, hydrodynamics and sediment transport were investigated. The analysis and findings are important contributions to delta management and planning, particularly contributing to the delta master plan. Nevertheless, this thesis did not include scenarios of delta planning (such as land use and delta-base water infrastructure planning) and climate change, suggesting recommendations for future studies. The validated model is an efficient tool to investigate possible changes of sediment transport in the Mekong Delta in the context of climate change and sea level rise. The climate change in the Mekong Basin may change the hydrograph at the input station (Kratie) of the Mekong Delta, ranging from -10 % to + 15 % in the period (2032-2042) compared to the period (1982-1992) (Lauri et al., 2012). Additionally, hydropower dams highly drive seasonal variations of flows in the Mekong River. Impacts of these factors on the Mekong Delta can be explored by modifying boundary conditions at Kratie in the projected range.

REFERENCES

Abily, M., Bertrand, N., Delestre, O., Gourbesville, P. and Duluc, C. M.: Spatial Global Sensitivity Analysis of High Resolution classified topographic data use in 2D urban flood modelling, Environ. Model. Softw., 77, 183–195, doi:10.1016/j.envsoft.2015.12.002, 2016.

Achete, F. M., van der Wegen, M., Roelvink, D. and Jaffe, B.: A 2-D process-based model for suspended sediment dynamics: a first step towards ecological modeling, Hydrol. Earth Syst. Sci., 19(6), 2837–2857, doi:10.5194/hess-19-2837-2015, 2015.

Achete, F. M., van der Wegen, M., Roelvink, D. and Jaffe, B.: Suspended sediment dynamics in a tidal channel network under peak river flow, Ocean Dyn., 66(5), 703–718, doi:10.1007/s10236-016-0944-0, 2016.

ADB: Climate Risks in the Mekong Delta: Ca Mau and Kien Giang Provinces of Viet Nam., 2013.

Amante, C. and Eakins, B. W.: ETOPO1 1 arc-minute global relief model: Procedures, data sources and analysis, NOAA Tech. Memo. NESDIS, NGDC-24(March), 19 pp, doi:10.1594/PANGAEA.769615, 2009.

Anthony, E. J., Brunier, G., Besset, M., Goichot, M., Dussouillez, P. and Nguyen, V. L.: Linking rapid erosion of the Mekong River Delta to human activities, Sci. Rep., 5, 14745, doi:10.1038/srep14745, 2015.

Bailly du Bois, P.: Automatic calculation of bathymetry for coastal hydrodynamic models, Comput. Geosci., 37(9), 1303–1310, doi:10.1016/j.cageo.2010.11.018, 2011.

Di Baldassarre, G. and Montanari, A.: Uncertainty in river discharge observations: A quantitative analysis, Hydrol. Earth Syst. Sci., 13(6), 913–921, doi:10.5194/hess-13-913-2009, 2009.

Bates, P. D., Horritt, M. S., Aronica, G. and Beven, K.: Bayesian updating of flood inundation likelihoods conditioned on flood extent data, Hydrol. Process., 18(17), 3347–3370, doi:10.1002/hyp.1499, 2004.

Berlamonta, J., Ockenden, M., Toormana, E. and Winterwerp, J.: The characterisation of cohesive sediment properties, Coast. Eng., 21, 105–128, 1993.

Biggs, D., Miller, F., Hoanh, C. T. and Molle, F.: The delta machine: water management in the Vietnamese Mekong Delta in historical and contemporary perspectives, in Contested waterscapes in the Mekong region: hydropower, livelihoods and governance, edited by F. Molle, T. Foran, and M. Kakonen, pp. 203–225, Routledge, London, UK., 2009.

Bomers, A., Schielen, R. M. J. and Hulscher, S. J. M. H.: The influence of grid shape and grid size on hydraulic river modelling performance, Environ. Fluid Mech., 19(5), 1273–1294, doi:10.1007/s10652-019-09670-4, 2019.

Booij, N., Ris, R. C. and Holthuijsen, L. H.: A third-generation wave model for coastal regions: 1. Model description and validation, J. Geophys. Res., 104(C4), 7649–7666, doi:10.1029/98JC02622, 1999.

de Boor, C.: A Practical Guide to Splines, edited by J. E. Marsden and L. Sirovich, Applied Mathematical Sciences., 2001.

Bravard, J.-P., Goichot, M. and Gaillot, S.: Geography of sand and gravel mining in the Lower Mekong River, EchoGéo, (26), 0–20, doi:10.4000/echogeo.13659, 2013.

Brunier, G., Anthony, E. J., Goichot, M., Provansal, M. and Dussouillez, P.: Recent morphological changes in the Mekong and Bassac river channels, Mekong delta: the marked impact of river-bed mining and implications for delta destabilisation, Geomorphology, 224, 177–191, doi:10.1016/j.geomorph.2014.07.009, 2014.

Carter, G. S. and Shankar, U.: Creating rectangular bathymetry grids for environmental numerical modelling of gravel-bed rivers, Appl. Math. Model., 21(11), 699–708, doi:10.1016/S0307-904X(97)00094-2, 1997.

Caviedes-Voullième, D., Morales-Hernández, M., López-Marijuan, I. and García-Navarro, P.: Reconstruction of 2D river beds by appropriate interpolation of 1D cross-sectional information for flood simulation, Environ. Model. Softw., 61(July 2015), 206–228, doi:10.1016/j.envsoft.2014.07.016, 2014.

Chapman, A. and Darby, S.: Evaluating sustainable adaptation strategies for vulnerable mega-deltas using system dynamics modelling: Rice agriculture in the Mekong Delta's An Giang Province, Vietnam, Sci. Total Environ., 559, 326–338, doi:10.1016/j.scitotenv.2016.02.162, 2016.

Chapman, A. D., Darby, S. E., Hoang, H. M., Tompkins, E. L. and Van, T. P. D.: Adaptation and development trade-offs: fluvial sediment deposition and the sustainability of rice-cropping in An Giang Province, Mekong Delta, Clim. Change, 137, 593–608, doi:10.1007/s10584-016-1684-3, 2016.

Chen, W. B. and Liu, W. C.: Modeling the influence of river cross-section data on a river stage using a two-dimensional/ three-dimensional hydrodynamic model, Water (Switzerland), 9(3), doi:10.3390/w9030203, 2017.

Conner, J. T. and Tonina, D.: Effect of cross-section interpolated bathymetry on 2D hydrodynamic model results in a large river, Earth Surf. Process. Landforms, 39(4), 463–475, doi:10.1002/esp.3458, 2014.

Cosslett, T. L. and Cosslett, P. D.: Water Resources and Food Security in the Vietnam Mekong Delta., 2014.

Curtarelli, M., Leão, J., Ogashawara, I., Lorenzzetti, J. and Stech, J.: Assessment of spatial interpolation methods to map the bathymetry of an Amazonian hydroelectric reservoir to aid in decision making for water management, ISPRS Int. J. Geo-Information, 4, 220–235, doi:10.3390/ijgi4010220, 2015.

Dang, D. T., Cochrane, T. A., Arias, M. E. and Dang, V. P.: Future hydrological alterations in the Mekong Delta under the impact of water resources development, land subsidence and sea level rise, J. Hydrol. Reg. Stud., 15(November 2017), 119–133, doi:10.1016/j.ejrh.2017.12.002, 2018a.

Dang, T. H., Ouillon, S. and Vinh, G. Van: Water and suspended sediment budgets in the Lower Mekong from high-frequency, Water, 10(846), doi:10.3390/w10070846, 2018b.

Darby, S. E., Hackney, C. R., Leyland, J., Kummu, M., Lauri, H., Parsons, D. R., Best, J. L., Nicholas, A. P. and Aalto, R.: Fluvial sediment supply to a mega-delta reduced by shifting tropical-cyclone activity, Nat. Publ. Gr., 539, 276–279, doi:10.1038/nature19809, 2016.

Dartmouth Flood Observatory: Vietnam and Cambodia Lower Mekong: Rapid response inundation map. [online] Available from: http://www.dartmouth.edu/~floods/Archives/2004sum.htm, 2004.

Dastgheib, A., Roelvink, J. A. and Wang, Z. B.: Long-term process-based morphological modeling of the Marsdiep Tidal Basin, Mar. Geol., 256(1–4), 90–100, doi:10.1016/j.margeo.2008.10.003, 2008.

Deltares: Delft3D-FLOW manual., 2014.

Deltares: D-Flow 1D (SOBEK 3): Hydrodynamics, Delft, The Netherlands., 2018.

Deltares: D-Flow Flexible Mesh: User Manual, Delft, the Netherlands., 2020a.

Deltares: Delft3D Flexible Mesh Suite: Technical reference manual., 2020b.

DeMaster, D. J., Liu, J. P., Eidam, E., Nittrouer, C. A. and Nguyen, T. T.: Determining rates of sediment accumulation on the Mekong shelf: timescales, steady-state assumptions, and radiochemical tracers, Cont. Shelf Res., 147, 182–196, 2017.

Department of Agriculture and Rural Development of An Giang and Dong Thap: Maps of hydraulic infrastructure in An Giang and Dong Thap, An Giang and Dong Thap., 2012.

Dilbone, E., Legleiter, C. J., Alexander, J. S. and McElroy, B.: Spectrally based bathymetric mapping of a dynamic, sand-bedded channel: Niobrara River, Nebraska, USA, River Res. Appl., 34(5), 430–441, doi:10.1002/rra.3270, 2018.

Dinh, Q., Balica, S., Popescu, I. and Jonoski, a.: Climate change impact on flood hazard, vulnerability and risk of the Long Xuyen Quadrangle in the Mekong Delta, Int. J. River Basin Manag., 10(1), 103–120, doi:10.1080/15715124.2012.663383, 2012.

Dissanayake, P. and Wurpts, A.: Modelling an anthropogenic effect of a tidal basin evolution applying tidal and wave boundary forcings: Ley Bay, East Frisian Wadden Sea, Coast. Eng., 82, 9–24, doi:10.1016/j.coastaleng.2013.08.005, 2013.

Dung, N. V., Merz, B., Bárdossy, a., Thang, T. D. and Apel, H.: Multi-objective automatic calibration of hydrodynamic models utilizing inundation maps and gauge data, Hydrol. Earth Syst. Sci., 15(4), 1339–1354, doi:10.5194/hess-15-1339-2011, 2011.

Duong, T. M., Ranasinghe, R., Luijendijk, A., Dastgheib, A. and Roelvink, D.: Climate change impacts on the stability of small tidal inlets: a numerical modelling study using the realistic analogue approach, Int. J. Ocean Clim. Syst., 3(October 2015), 163–172, doi:10.1260/1759-3131.3.3.163, 2012.

Duong, V. H. T., Nestmann, F., Van, T. C., Hinz, S., Oberle, P. and Geiger, H.: Geographical impact of dyke measurement for land use on flood water in geographical impact of dyke measurement for land use on flood water in the Mekong Delta, in 8th Eastern European Young Water Professionals Conference - IWA, pp. 308–317, Gdansk, Poland, 12-14 May, 2016., 2016.

Dutta, D., Alam, J., Umeda, K., Hayashi, M. and Hironaka, S.: A two-dimensional hydrodynamic model for flood inundation simulation: A case study in the lower Mekong river basin, Hydrol. Process., 21, 1223–1237, doi:10.1002/hyp, 2007.

Egbert, G. D. and Erofeeva, S. Y.: Efficient inverse modeling of barotropic ocean tides, J. Atmos. Ocean. Technol., 19(2), 183–204, 2002.

Eidam, E. F., Nittrouer, C. A., Ogston, A. S., DeMaster, D. J., Liu, J. P., Nguyen, T. T. and Nguyen, T. N.: Dynamic controls on shallow clinoform geometry: Mekong Delta, Vietnam, Cont. Shelf Res., 147, 165–181, 2017.

Ferré, B., Sherwood, C. R. and Wiberg, P. L.: Sediment transport on the Palos Verdes shelf, California, Cont. Shelf Res., 30(7), 761–780, doi:10.1016/j.csr.2010.01.011, 2010.

Frappart, F., Do Minh, K., L'Hermitte, J., Cazenave, A., Ramillien, G., Le Toan, T. and Mognard-Campbell, N.: Water volume change in the lower Mekong from satellite altimetry and imagery data, Geophys. J. Int., 167(2), 570–584, doi:10.1111/j.1365-246X.2006.03184.x, 2006.

Fujihara, Y., Hoshikawa, K., Fujii, H., Kotera, A., Nagano, T. and Yokoyama, S.: Analysis and attribution of trends in water levels in the Vietnamese Mekong Delta, Hydrol. Process., 30(6), 835–845, doi:10.1002/hyp.10642, 2016.

Fujii, H., Garsdal, H., Ward, P., Ishii, M., Morishita, K. and Boivin, T.: Hydrological roles of the Cambodian floodplain of the Mekong River, Int. J. River Basin Manag., 1(3), 1–14, doi:10.1080/15715124.2003.9635211, 2003.

Furukawa, K. and Wolanski, E.: Sedimentation in mangrove forests, Mangroves Salt Marshes, (1), 3–10, doi:10.1023/A:1025973426404, 1996.

Furukawa, K., Wolanski, E. and Mueller, H.: Currents and Sediment Transport in Mangrove Forests, Estuar. Coast. Shelf Sci., 44(3), 301–310, doi:10.1006/ecss.1996.0120, 1997.

Garcia, M., Ramirez, I., Verlaan, M. and Castillo, J.: Application of a three-dimensional hydrodynamic model for San Quintin Bay, B.C., Mexico. validation and calibration using OpenDA, J. Comput. Appl. Math., 273, 428–437, doi:10.1016/j.cam.2014.05.003, 2015.

Gebrehiwot, K. A., Haile, A. M., de Fraiture, C. M. S., Chukalla, A. D. and Embaye, T. G.: Optimizing flood and sediment management of spate irrigation in Aba'ala Plains, Water Resour. Manag., 29(3), 833–847, doi:10.1007/s11269-014-0846-1, 2014.

Gibson, S., Comport, B. and Corum, Z.: Calibrating a sediment transport model through a gravel-sand transition: Avoiding equifinality errors in HEC-RAS models of the Puyallup and White rivers, in World Environmental and Water Resources Congress 2017, edited by C. N. Dunn and B. Van Weele, pp. 179–191, Sacramento, California., 2017.

Goff, J. A. and Nordfjord, S.: Interpolation of fluvial morphology using channel-oriented coordinate transformation: A case study from the New Jersey Shelf, Math. Geol., 36(6), 643–658, doi:10.1023/B:MATG.0000039539.84158.cd, 2004.

Gratiot, N., Bildstein, A., Anh, T. T., Thoss, H., Denis, H., Michallet, H. and Apel, H.: Sediment flocculation in the Mekong River estuary, Vietnam, an important driver of geomorphological changes, Comptes Rendus - Geosci., 349(6–7), 260–268, doi:10.1016/j.crte.2017.09.012, 2017.

GSOVN, (GENERAL STATISTICS OFFICE of VIET NAM): Statistical yearbook of Vietnam, Hanoi., 2010.

Gugliotta, M., Saito, Y., Van Lap, N., Thi Kim Oanh, T., Nakashima, R., Tamura, T., Uehara, K., Katsuki, K. and Yamamoto, S.: Process regime, salinity, morphological, and sedimentary trends along the fluvial to marine transition zone of the mixed-energy Mekong River delta, Vietnam, Cont. Shelf Res., 147(August), 7–26, 2017.

Guo, L., van der Wegen, M., Wang, Z. B., Roelvink, D. and He, Q.: Exploring the impacts of multiple tidal constituents and varying river flow on long-term, large-scale estuarine morphodynamics by means of a 1-D model, J. Geophys. Res. Earth Surf., 121, 1000–1022, doi:10.1002/2016JF003821, 2016.

Gupta, A. and Liew, S. C.: The Mekong from satellite imagery: A quick look at a large river, Geomorphology, 85(3–4), 259–274, doi:10.1016/j.geomorph.2006.03.036, 2007.

Gupta, A., Liew, S. C. and Heng, A. W. C.: Sediment storage and transfer in the Mekong: Generalizations on a large river, IAHS-AISH Publ., (306), 450–459 [online] Available from: http://www.scopus.com/inward/record.url?eid=2-s2.0-33748435041&partnerID=tZOtx3y1, 2006.

Hasan, G. M. J., van Maren, D. S. and Fatt, C. H.: Numerical study on mixing and stratification in the ebb-dominant Johor estuary, J. Coast. Res., 29(2), 201–215, doi:10.2112/JCOASTRES-D-12-00053.1, 2012.

Hawker, L., Rougier, J., Neal, J., Bates, P., Archer, L. and Yamazaki, D.: Implications of simulating global digital elevation models for flood inundation studies, Water Resour. Res., 54, 7910–7928, doi:10.1029/2018WR023279, 2018.

Heege, T., Kiselev, V., Wettle, M. and Hung, N. N.: Operational multi-sensor monitoring of turbidity for the entire Mekong Delta, Int. J. Remote Sens., 35(8), 2910–2926, doi:10.1080/01431161.2014.890300, 2014.

Hein, H., Hein, B. and Pohlmann, T.: Recent sediment dynamics in the region of Mekong water influence, Glob. Planet. Change, 110, 183–194, doi:10.1016/j.gloplacha.2013.09.008, 2013.

Hilton, J. E., Grimaldi, S., Cohen, R. C. Z., Garg, N., Li, Y., Marvanek, S., Pauwels, V. R. N. and Walker, J. P.: River reconstruction using a conformal mapping method, Environ. Model. Softw., 119(June), 197–213, doi:10.1016/j.envsoft.2019.06.006, 2019.

Hoanh, C. T., Phong, N. D., Gowing, J. W., Tuong, T. P., Ngoc, N. V. and Hien, N. X.: Hydraulic and water quality modeling: A tool for managing land use conflicts in inland coastal zones, Water Policy, 11(SUPPL. 1), 106–120, doi:10.2166/wp.2009.107, 2009.

Hu, K., Ding, P., Wang, Z. and Yang, S.: A 2D/3D hydrodynamic and sediment transport model for the Yangtze estuary, China, J. Mar. Syst., 77(1–2), 114–136, doi:10.1016/j.jmarsys.2008.11.014, 2009.

Huang, C. L., Wang, H. W. and Hou, J. L.: Estimating spatial distribution of daily snow depth with kriging methods: combination of MODIS snow cover area data and ground-based observations, Cryosph. Discuss., 9(5), 4997–5020, doi:10.5194/tcd-9-4997-2015,

2015.

Hung, N. N.: Sediment dynamics in the floodplain of the Mekong Delta, Vietnam, Universität Stuttgart., 2011.

Hung, N. N., Delgado, J. M., Güntner, A., Merz, B., Bárdossy, A. and Apel, H.: Sedimentation in the floodplains of the Mekong Delta, Vietnam. Part I: suspended sediment dynamics, Hydrol. Process., 28(May 2013), 3132–3144, doi:10.1002/hyp.9856, 2014a.

Hung, N. N., Delgado, J. M., Güntner, A., Merz, B., Bárdossy, A. and Apel, H.: Sedimentation in the floodplains of the Mekong Delta, Vietnam Part II: deposition and erosion, Hydrol. Process., 28(7), 3145–3160, doi:10.1002/hyp.9855, 2014b.

IPCC: Climate Change 2007: Impacts, adaptation and vulnerability: contribution of Working Group II to the fourth assessment report of the Intergovernmental Panel, edited by M. L. Parry, O. F. Canziani, J. P. Palutikof, P. J. van der Linden, and C. E. Hanson, Cambridge University Press, Cambridge, UK., 2007.

Ji, Z.-G.: Hydrodynamics and Water Quality: Modeling Rivers, Lakes and Estuaries, JOHN WILEY & SONS, INC., 2008.

Ji, Z.: Water Quality and Eutrophication, in Hydrodynamics and Water Quality: Modeling Rivers, Lakes, and Estuaries, John Wiley & Sons Inc., 2017.

Kakonen, M.: Mekong Delta at the crossroads: More control or adaptation?, Ambio, 37(3), 205–212, 2008.

Karamouz, M., Nazif, S. and Falahi, M.: Hydrology and Hydroclimatology: Principles and Applications., 2013.

Kernkamp, H. W. J., Van Dam, A., Stelling, G. S. and De Goede, E. D.: Efficient scheme for the shallow water equations on unstructured grids with application to the continental shelf, Ocean Dyn., 61(8), 1175–1188, doi:10.1007/s10236-011-0423-6, 2011.

van Kessel, T., Vanlede, J. and de Kok, J.: Development of a mud transport model for the Scheldt estuary, Cont. Shelf Res., 31, S165–S181, doi:10.1016/j.csr.2010.12.006, 2011.

Kite, G.: Modelling the Mekong: hydrological simulation for environmental impact studies, J. Hydrol., 253, 1–13, doi:10.1016/S0022-1694(01)00396-1, 2001.

Koehnken, L.: IKMP discharge and dediment monitoring program review, recommendations and data analysis. Part 2: Data analysis of preliminary results, Mekong River Commission, Phnom Penh, Cambodia., 2012.

Koehnken, L.: Discharge sediment monitoring project (DSMP) 2009 – 2013 summary & analysis of results., 2014.

Kondolf, G. M., Schmitt, R. J. P., Carling, P., Darby, S., Arias, M., Bizzi, S., Castelletti, A., Cochrane, T. A., Gibson, S., Kummu, M., Oeurng, C., Rubin, Z. and Wild, T.: Changing sediment budget of the Mekong: Cumulative threats and management strategies for a large river basin, Sci. Total Environ., 625, 114–134, doi:10.1016/j.scitotenv.2017.11.361, 2018.

Kuang, C., Chen, W., Gu, J., Su, T. C., Song, H., Ma, Y. and Dong, Z.: River discharge contribution to sea-level rise in the Yangtze River Estuary, China, Cont. Shelf Res., 134(June 2016), 63–75, doi:10.1016/j.csr.2017.01.004, 2017.

Kuenzer, C., Guo, H., Huth, J., Leinenkugel, P., Li, X. and Dech, S.: Flood mapping and flood dynamics of the mekong delta: ENVISAT-ASAR-WSM based time series analyses, Remote Sens., 5(2), 687–715, doi:10.3390/rs5020687, 2013.

Kummu, M. and Varis, O.: Sediment-related impacts due to upstream reservoir trapping, the Lower Mekong River, Geomorphology, 85(3–4), 275–293, doi:10.1016/j.geomorph.2006.03.024, 2007.

Kummu, M., Penny, D., Sarkkula, J. and Koponen, J.: Sediment: curse or blessing for Tonle Sap Lake?, Ambio, 37(3), 158–63, doi:10.1579/0044-7447(2008)37[158:scobft]2.0.co;2, 2008.

Kummu, M., Tes, S., Yin, S., Adamson, P., Józsa, J., Koponen, J., Richey, J. and Sarkkula, J.: Water balance analysis for the Tonle Sap Lake-floodplain system, Hydrol. Process., 28(4), 1722–1733, doi:10.1002/hyp.9718, 2014a.

Kummu, M., Tes, S., Yin, S., Adamson, P., J??zsa, J., Koponen, J., Richey, J. and Sarkkula, J.: Water balance analysis for the Tonle Sap Lake-floodplain system, Hydrol. Process., 28(4), 1722–1733, doi:10.1002/hyp.9718, 2014b.

Lai, R., Wang, M., Yang, M. and Zhang, C.: Method based on the Laplace equations to reconstruct the river terrain for two-dimensional hydrodynamic numerical modeling, Comput. Geosci., 111, 26–38, doi:10.1016/j.cageo.2017.10.006, 2018.

Landers, M. N. and Sturm, T. W.: Hysteresis in suspended sediment to turbidity relations due to changing particle size distributions, Water Resour. Res., 49, 5487–5500, 2013.

Lauri, H., de Moel, H., Ward, P. J., Räsänen, T. a., Keskinen, M. and Kummu, M.: Future changes in Mekong River hydrology: impact of climate change and reservoir operation on discharge, Hydrol. Earth Syst. Sci., 16(12), 4603–4619, doi:10.5194/hess-16-4603-2012, 2012.

Le, H.-A., Gratiot, N., Santini, W., Ribolzi, O., Soares-Frazao, S. and Deleersnijder, E.: Sediment properties in the fluvial and estuarine environments of the Mekong River, in E3S Web of Conferences, vol. 40, p. 05063., 2018.

Lê Sâm: Thủy nông ở đồng bằng sông Cửu Long, Nhà xuất bản Nông Nghiệp, Ho Chi Minh city., 1996.

Le, T. V. H., Nguyen, H. N., Wolanski, E., Tran, T. C. and Haruyama, S.: The combined impact on the flooding in Vietnam's Mekong River delta of local man-made structures, sea level rise, and dams upstream in the river catchment, Estuar. Coast. Shelf Sci., 71(1–2), 110–116, doi:10.1016/j.ecss.2006.08.021, 2007.

Le, T. V. H., Shigeko, H., Nhan, N. H. and Cong, T. T.: Infrastructure effects on floods in the Mekong River Delta in Vietnam, Hydrol. Process., 22, 1359–1372, doi:10.1002/hyp, 2008.

Legleiter, C. J.: Mapping river depth from publicly available aerial images, River Res. Appl., 29, 760–780, doi:10.1002/rra, 2013.

Legleiter, C. J. and Kyriakidis, P. C.: Spatial prediction of river channel topography by kriging, Earth Surf. Process. Landforms, 33(6), 841–967, doi:10.1002/esp, 2008.

Lesser, G. R., Roelvink, J. A., van Kester, J. A. T. M. and Stelling, G. S.: Development and validation of a three-dimensional morphological model, Coast. Eng., 51(8–9), 883–915, doi:10.1016/j.coastaleng.2004.07.014, 2004.

Li, J. and Heap, A. D.: A review of comparative studies of spatial interpolation methods in environmental sciences: Performance and impact factors, Ecol. Inform., 6, 228–241, doi:10.1016/j.ecoinf.2010.12.003, 2011.

Van Liew, M. W., Veith, T. L., Bosch, D. D. and Arnold, J. G.: Suitability of SWAT for the conservation effects assessment project: Comparison on USDA agricultural research service watersheds, J. Hydrol. Eng., 12(2), 173–189, doi:10.1016/j.lwt.2017.04.076, 2007.

Lin, G. and Chen, L.: A spatial interpolation method based on radial basis function networks incorporating a semivariogram model, J. Hydrol., 288, 288–298, doi:10.1016/j.jhydrol.2003.10.008, 2004.

Loisel, H., Mangin, A., Vantrepotte, V., Dessailly, D., Dinh, D. N., Garnesson, P., Ouillon, S., Lefebvre, J., Mériaux, X. and Phan, T. M.: Remote sensing of environment variability of suspended particulate matter concentration in coastal waters under the Mekong's influence from ocean color (MERIS) remote sensing over the last decade, Remote Sens. Environ., 150, 218–230, doi:10.1016/j.rse.2014.05.006, 2014.

Loucks, D. P.: River models, in Encyclopedia of Ecology, pp. 3069–3083., 2008.

Lu, S., Tong, C., Lee, D.-Y., Zheng, J., Shen, J., Zhang, W. and Yan, Y.: Propagation of tidal waves up in Yangtze estuary during the dry season, J. Geophys. Res. Ocean., 120, doi:10.1002/2014JC010632, 2015.

Lu, X., Kummu, M. and Oeurng, C.: Reappraisal of sediment dynamics in the Lower Mekong River, Cambodia, Earth Surf. Process. Landforms, 39(14), 1855–1865, doi:10.1002/esp.3573, 2014.

Lu, X. X. and Siew, R. Y.: Water discharge and sediment flux changes in the Lower Mekong River, Hydrol. Earth Syst. Sci., 10, 181–195, doi:10.5194/hessd-2-2287-2005, 2006.

Manh, N. V., Merz, B. and Apel, H.: Sedimentation monitoring including uncertainty analysis in complex floodplains: A case study in the Mekong Delta, Hydrol. Earth Syst. Sci., 17(8), 3039–3057, doi:10.5194/hess-17-3039-2013, 2013.

Manh, N. V., Dung, N. V., Hung, N. N., Merz, B. and Apel, H.: Large-scale quantification of suspended sediment transport and deposition in the Mekong Delta, Hydrol. Earth Syst. Sci., 18, 3033–3053, doi:10.5194/hessd-11-4311-2014, 2014.

Manh, N. Van, Dung, N. V., Hung, N. N., Kummu, M., Merz, B. and Apel, H.: Future sediment dynamics in the Mekong Delta floodplains: impacts of hydropower development, climate change and sea level rise, Glob. Planet. Change, 127, 22–33, doi:10.1016/j.gloplacha.2015.01.001, 2015.

Marchesiello, P., Nguyen, N. M., Gratiot, N., Loisel, H., Anthony, E. J., Dinh, C. S., Nguyen, T., Almar, R. and Kestenare, E.: Erosion of the coastal Mekong delta: Assessing natural against man induced processes, Cont. Shelf Res., 181(November 2018), 72–89, doi:10.1016/j.csr.2019.05.004, 2019.

van Maren, D. S. and Cronin, K.: Uncertainty in complex three-dimensional sediment transport models: equifinality in a model application of the Ems Estuary, the Netherlands, Ocean Dyn., 66(12), 1665–1679, doi:10.1007/s10236-016-1000-9, 2016.

Marineau, M. D. and Wright, S. A.: Effects of human alterations on the hydrodynamics and sediment transport in the Sacramento-San Joaquin Delta, California, in IAHS-AISH Proceedings and Reports, vol. 367, pp. 399–406., 2014.

Martyr-Koller, R. C., Kernkamp, H., van Dam, A., van der Wegen, M., Lucas, L. V., Knowles, N., Jaffee, B. and Fregoso, T. A.: Application of an unstructured 3D finite volume numerical model for hydrodynamic and water-quality transport in the San Francisco Bay-Delta system, Estuar. Coast. Shelf Sci., 192, 86–107, doi:10.1016/j.ecss.2017.04.024, 2017.

Mazda, Y., Kanazawa, N. and Wolanski, E.: Tidal asymmetry in mangrove creeks, Hydrobiologia, 295(1–3), 51–58, doi:10.1007/BF00029110, 1995.

Mazda, Y., Magi, M., Ikeda, Y., Kurokawa, T. and Asano, T.: Wave reduction in a mangrove forest dominated by Sonneratia sp., Wetl. Ecol. Manag., 14(4), 365–378, doi:10.1007/s11273-005-5388-0, 2006.

McLachlan, R., Ogston, A. and Allison, M.: Implications of tidally varying bed shear stress and intermittent estuarine stratification on fine-sediment dynamics through the Mekong's tidal river to estuarine reach, Cont. Shelf Res., 147, 27–37, 2017.

Mclachlan, R. L., Ogston, A. S. and Allison, M. A.: Implications of tidally-varying bed stress and intermittent estuarine stratification on fine-sediment dynamics through the Mekong's tidal river to estuarine reach, Cont. Shelf Res., 147, 27–37, doi:10.1016/j.csr.2017.07.014, 2017.

Merwade, V.: Effect of spatial trends on interpolation of river bathymetry, J. Hydrol., 371, 169–181, doi:10.1016/j.jhydrol.2009.03.026, 2009.

Merwade, V., Cook, A. and Coonrod, J.: GIS techniques for creating river terrain models for hydrodynamic modeling and flood inundation mapping, Environ. Model. Softw., 23(10–11), 1300–1311, doi:10.1016/j.envsoft.2008.03.005, 2008.

Mhashhash, A., Bockelmann-Evans, B. and Pan, S.: Effect of hydrodynamics factors on sediment flocculation processes in estuaries, J. Soils Sediments, 18, 3094–3103, doi:10.1007/s11368-017-1837-7, 2018.

Milliman, J. D. and Farnsworth, K. L.: River Discharge to the Coastal Ocean. A Global Synthesis, Cambridge University Press, the UK., 2011.

Milliman, J. D. and Syvitski, J. P. M.: Geomorphic/tectonic control of sediment discharge to the ocean: the importance of small mountainous rivers, J. Geol., 100(5), 525–544, doi:10.1086/629606, 1992.

Mitas, L. and Mitasova, H.: Spatial interpolation, in Geographic Information Systems: Principles, Techniques, Management and Applications, pp. 481–492., 2005.

Moriasi, D. N., J. G. Arnold, M. W. Van Liew, R. L. Bingner, R. D. Harmel and T. L. Veith: Model Evaluation Guidelines for Systematic Quantification of Accuracy in

Watershed Simulations, Trans. ASABE, 50(3), 885–900, doi:10.13031/2013.23153, 2007.

MRC: Overview of the Hydrology of the Mekong Basin, Vientiane, Laos., 2005.

MRC: Annual Mekong flood report 2006., 2007.

MRC: Annual Mekong Flood Report 2008, Vientiane., 2009a.

MRC: MRC Management Information Booklet No.2: The Flow of the Mekong. [online] Available from: http://www.mrcmekong.org/assets/Publications/report-management-develop/MRC-IM-No2-the-flow-of-the-mekong.pdf (last access: 15/04/2014), 2009b.

MRC: State of the Basin Report 2010, Vientiane, Laos., 2010.

MRC: Planning Atlas of the Lower Mekong River Basin. [online] Available from: http://www.mrcmekong.org/assets/Publications/basin-reports/BDP-Atlas-Final-2011.pdf, 2011.

MRC: Weekly flood situation report for the Mekong River Basin., 2016.

Mullarney, J. C., Henderson, S. M., Reyns, J. A. H., Norris, B. K. and Bryan, K. R.: Spatially varying drag within a wave-exposed mangrove forest and on the adjacent tidal flat, Cont. Shelf Res., 147(June), 102–113, doi:10.1016/j.csr.2017.06.019, 2017.

Nash, J. E. and Sutcliffe, J. V: River flow forecasting through conceptual models Part I-a discussion of principles*, J. Hydrol., 10, 282–290, doi:10.1016/0022-1694(70)90255-6, 1970.

Nguyen, a. D. and Savenije, H. H. G.: Salt intrusion in multi-channel estuaries: a case study in the Mekong Delta, Vietnam, Hydrol. Earth Syst. Sci. Discuss., 3(2), 499–527, doi:10.5194/hessd-3-499-2006, 2006.

Nguyen, A. D., Savenije, H. H. G., Pham, D. N. and Tang, D. T.: Using salt intrusion measurements to determine the freshwater discharge distribution over the branches of a multi-channel estuary: The Mekong Delta case, Estuar. Coast. Shelf Sci., 77(3), 433–445, doi:10.1016/j.ecss.2007.10.010, 2008.

Nguyen Anh Duc: Salt Intrusion, Tides and Mixing in Multi-channel Estuaries, Taylor & Francis, November., 2008.

Nguyen, P. M., Le, K. Van, Botula, Y. and Cornelis, W. M.: Evaluation of soil water retention pedotransfer functions for Vietnamese Mekong Delta soils, Agric. Water Manag., 158, 126–138, doi:10.1016/j.agwat.2015.04.011, 2015.

Nguyen Van Manh: Large-scale floodplain sediment dynamics in the Mekong Delta: present state and future prospects., 2014.

Nowacki, D. J., Ogston, A. S., Nittrouer, C. A., Fricke, A. T. and Van, D. T. P.: Sediment dynamics in the lower Mekong River: transition from tidal river to estuary, J. Geophys. Res. Ocean., 120, doi:10.1002/2014JC010632, 2015.

Ogston, A. S., Allison, M. A., Mullarney, J. C. and Nittrouer, C. A.: Sediment- and hydro-dynamics of the Mekong Delta: From tidal river to continental shelf, Cont. Shelf Res., 147(September), 1–6, doi:10.1016/j.csr.2017.08.022, 2017.

Overeem, I. and Syvitski, J. P. M.: Dynamics and vulnerability of delta systems., 2009.

Overeem, I., Kettner, A. J. and Syvitski, J. P. M.: Impacts of humans on river fluxes and morphology, in Treatise on Geomorphology, vol. 9, edited by J. F. Shroder and E. Wohl, pp. 828–841, Academic Press., 2013.

Partheniades, E.: Erosion and deposition of cohesive soils, J. Hydraul. Div., 91(1), 105–139, 1965.

Pawlowicz, R., Beardsley, B. and Lentz, S.: Classical tidal harmonic analysis including error estimates in MATLAB using T_TIDE, Comput. Geosci., 28(8), 929–937, doi:10.1016/S0098-3004(02)00013-4, 2002.

Pebesma, E. J.: Multivariable geostatistics in S: The gstat package, Comput. Geosci., 30(7), 683–691, doi:10.1016/j.cageo.2004.03.012, 2004.

Portela, L. I., Ramos, S. and Teixeira, A. T.: Effect of salinity on the settling velocity of fine sediments of a harbour basin, J. Coast. Res., 1(65), 1188–1193, doi:10.2112/SI65-201, 2013.

Quartel, S., Kroon, A., Augustinus, P. G. E. F., Van Santen, P. and Tri, N. H.: Wave attenuation in coastal mangroves in the Red River Delta, Vietnam, J. Asian Earth Sci., 29(4), 576–584, doi:10.1016/j.jseaes.2006.05.008, 2007.

Renaud, F. G. and Kuenzer, C.: The Mekong Delta System, Springer, Dordrecht, Heidelberg, New York, London., 2012.

Renaud, F. G., Syvitski, J. P. M., Sebesvari, Z., Werners, S. E., Kremer, H., Kuenzer, C., Ramesh, R., Jeuken, A. D. and Friedrich, J.: Tipping from the Holocene to the Anthropocene: How threatened are major world deltas?, Curr. Opin. Environ. Sustain., 5(6), 644–654, doi:10.1016/j.cosust.2013.11.007, 2013.

Reuter, H. I., Nelson, A. and Jarvis, A.: An evaluation of void-filling interpolation methods for SRTM data, Int. J. Geogr. Inf. Sci., 21(9), 983–1008, doi:10.1080/13658810601169899, 2007.

Rivest, M., Marcotte, D. and Pasquier, P.: Hydraulic head field estimation using kriging with an external drift: A way to consider conceptual model information, J. Hydrol., 361(3–4), 349–361, doi:10.1016/j.jhydrol.2008.08.006, 2008.

Roelvink, D. and Walstra, D.-J.: Keep it simple by using complex models, Adv. hydro-science -engineering, VI, 1–11, 2004.

RoyalHaskoningHDV, Deltares and UNESCO-IHE: The flood management and mitigation programme. Component 2: Structural measures and flood proofing in the Lower Mekong Basin., 2010.

Sarkkula, J., Koponen, J., Lauri, H. and Virtanen, M.: Origin, fate and impacts of the Mekong sediments., 2010.

Savage, J. T. S., Bates, P., Freer, J., Neal, J. and Aronica, G.: When does spatial resolution become spurious in probabilistic flood inundation predictions?, Hydrol. Process., 30(13), 2014–2032, doi:10.1002/hyp.10749, 2016.

Sear, D. A. and Milne, J. A.: Surface modelling of upland river channel topography and sedimentology using GIS, Phys. Chem. Earth, Part B Hydrol. Ocean. Atmos., 25(4), 399–406, doi:10.1016/S1464-1909(00)00033-2, 2000.

Stephens, J. D., Allison, M. A., Leonardo, D. R. Di, Weathers, H. D., Ogston, A. S., McLachlan, R. L., Xing, F. and Meselhe, E. A.: Sand dynamics in the Mekong River channel and export to the coastal ocean, Cont. Shelf Res., 147, 38–50, 2017.

Sutherland, J., Walstra, D. J. R., Chesher, T. J., van Rijn, L. C. and Southgate, H. N.: Evaluation of coastal area modelling systems at an estuary mouth, Coast. Eng., 51(2), 119–142, doi:10.1016/j.coastaleng.2003.12.003, 2004.

Syvitski, J. P. M. and Kettner, A.: Sediment flux and the Anthropocene, Philos. Trans. R. Soc. A Math. Phys. Eng. Sci., 369(1938), 957–975, doi:10.1098/rsta.2010.0329, 2011.

Syvitski, J. P. M., Kettner, A. J., Overeem, I., Hutton, E. W. H., Hannon, M. T., Brakenridge, G. R., Day, J., Vörösmarty, C., Saito, Y., Giosan, L. and Nicholls, R. J.: Sinking deltas due to human activities, Nat. Geosci., 2(10), 681–686, doi:10.1038/ngeo629, 2009.

Szczuciński, W., Jagodziński, R., Hanebuth, T. J. J., Stattegger, K., Wetzel, A., Mitręga, M., Unverricht, D. and Van Phach, P.: Modern sedimentation and sediment dispersal pattern on the continental shelf off the Mekong River delta, South China sea, Glob. Planet. Change, 110, 195–213, doi:10.1016/j.gloplacha.2013.08.019, 2013.

Ta, T. K. O., Nguyen, V. L., Tateishi, M., Kobayashi, I., Saito, Y. and Nakamura, T.: Sediment facies and Late Holocene progradation of the Mekong River Delta in Bentre Province, southern Vietnam: An example of evolution from a tide-dominated to a tide- and wave-dominated delta, Sediment. Geol., 152(3–4), 313–325, doi:10.1016/S0037-0738(02)00098-2, 2002.

Talley, L. D., Pickard, G. L., Emery, W. J. and Swift, J. H.: Dynamical Processes for Descriptive Ocean Circulation, in Descriptive Physical Oceanography: An Introduction, pp. 187–221., 2011.

Tarekegn, T. H. and Sayama, T.: Correction of SRTM dem artefacts by Fourier transform for flood inundation modeling, J. Japan Soc. Civ. Eng. Ser. B1 (Hydraulic Eng., 69(4), 193–198, 2013.

Teng, J., Jakeman, A. J., Vaze, J., Croke, B. F. W., Dutta, D. and Kim, S.: Flood inundation modelling: A review of methods, recent advances and uncertainty analysis, Environ. Model. Softw., 90, 201–216, doi:10.1016/j.envsoft.2017.01.006, 2017.

Thanh, V. Q., Hoanh, C. T., Trung, N. H. and Tri, V. P. D.: A bias-correction method of precipitation data generated by regional climate model, in Proceedings of International Symposium on GeoInformatics for Spatial-Infrastructure Development in Earth and Allied Sciences, pp. 182–187., 2014.

Thanh, V. Q., Reyns, J., Wackerman, C., Eidam, E. F. and Roelvink, D.: Modelling suspended sediment dynamics on the subaqueous delta of the Mekong River, Cont. Shelf Res., 147(August), 213–230, doi:10.1016/j.csr.2017.07.013, 2017.

Thanh, V. Q., Roelvink, D., van Der Wegen, M., Reyns, J., Kernkamp, H., Vinh, G. Van and Linh, V. T. P.: Flooding in the Mekong Delta: the impact of dyke systems on downstream hydrodynamics, Hydrol. Earth Syst. Sci., 24, 189–212, 2020a.

Thanh, V. Q., Roelvink, D., van der Wegen, M., Tu, L. X., Reyns, J. and Linh, V. T. P.: Spatial topographic interpolation for meandering channels, J. Waterw. Port, Coastal,

Ocean Eng., 146(5), 04020024, 2020b.

Tran, D. D., van Halsema, G., Hellegers, P. J. G. J., Phi Hoang, L., Quang Tran, T., Kummu, M. and Ludwig, F.: Assessing impacts of dike construction on the flood dynamics in the Mekong Delta, Hydrol. Earth Syst. Sci. Discuss., 22, 1875–1896, doi:10.5194/hess-2017-141, 2018.

Tran, T.: Transboundary Mekong River Delta (Cambodia and Vietnam), in The Wetland Book: II: Distribution, Description and Conservation, edited by C. M. Finlayson, G. R. Milton, R. C. Prentice, and N. C. Davidson, pp. 1–12, Springer Netherlands, Dordrecht., 2016.

Tri, V. P. D., Trung, N. H. and Tuu, N. T.: Flow dynamics in the Long Xuyen Quadrangle under the impacts of full-dyke systems and sea level rise, VNU J. Sci. earth Sci., 28(January), 205–214, 2012.

Tri, V. P. D., Trung, N. H. and Thanh, V. Q.: Vulnerability to flood in the Vietnamese Mekong Delta : mapping and uncertainty assessment, J. Environ. Sci. Eng. B, 2(April), 229–237, 2013.

Triet, N. V. K., Dung, N. V., Fujii, H., Kummu, M., Merz, B. and Apel, H.: Has dyke development in the Vietnamese Mekong Delta shifted flood hazard downstream?, Hydrol. Earth Syst. Sci., 21(8), 3991–4010, doi:10.5194/hess-21-3991-2017, 2017.

Tu, L. X., Thanh, V. Q., Reyns, J., Van, S. P., Anh, D. T., Dang, T. D. and Roelvink, D.: Sediment transport and morphodynamical modeling on the estuaries and coastal zone of the Vietnamese Mekong Delta, Cont. Shelf Res., 186, 64–76, doi:10.1016/j.csr.2019.07.015, 2019.

Unverricht, D., Szczuciński, W., Stattegger, K., Jagodziński, R., Le, X. T. and Kwong, L. L. W.: Modern sedimentation and morphology of the subaqueous Mekong Delta, Southern Vietnam, Glob. Planet. Change, 110, 223–235, doi:10.1016/j.gloplacha.2012.12.009, 2013.

Van, P. D. T., Popescu, I., Van Griensven, A., Solomatine, D. P., Trung, N. H. and Green, A.: A study of the climate change impacts on fluvial flood propagation in the Vietnamese Mekong Delta, Hydrol. Earth Syst. Sci., 16(12), 4637–4649, doi:10.5194/hess-16-4637-2012, 2012.

Vinh, V. D., Ouillon, S., Van Thao, N. and Ngoc Tien, N.: Numerical simulations of suspended sediment dynamics due to seasonal forcing in the Mekong coastal area, Water, 8(6), 255, doi:10.3390/w8060255, 2016.

Vo, K. T.: Hydrology and Hydraulic Infrastructure Systems in the Mekong Delta, Vietnam, in The Mekong Delta System—Interdisciplinary Analyses of a River Delta, edited by C. Renaud, F., Kuenzer, pp. 49–83, Springer: Heidelberg, Germany., 2012.

Vo Luong, H. P.: Bathymetric survey of the mouth of Song Hau, , Pers. comm., 2016.

Vu Duy Vinh, Tran Anh Tu, Tran Dinh Lan and Nguyen Ngoc Tien: Characteristics of suspended particulate matter and the coastal turbidity maximum areas of the Mekong River, J. Environ. Sci. Eng. A, 4(2), doi:10.17265/2162-5298/2015.02.002, 2015.

Wackerman, C., Hayden, A. and Jonik, J.: Deriving spatial and temporal context for point measurements of suspended sediment concentration using remote sensing imagery in the Mekong Delta, Cont. Shelf Res., 147, 231–245, 2017.

Walling, D. E.: The changing sediment load of the Mekong River, Ambio, 37(3), 150–157, 2008.

Walling, D. E.: The sediment load of the Mekong River, in The Mekong: Biophysical Environment of an International River Basin, pp. 113–142., 2009.

Walther, B. A. and Moore, J. L.: The concepts of bias, precision and accuracy, and their use in testing the performance of species richness estimators, with a literature review of estimator performance, Ecography (Cop.)., 28, 815–829, 2005.

Warner, J. C., Armstrong, B., He, R. and Zambon, J. B.: Development of a Coupled Ocean-Atmosphere-Wave-Sediment Transport (COAWST) modeling system, Ocean Model., 35(3), 230–244, doi:10.1016/j.ocemod.2010.07.010, 2010.

Wassmann, R., Hien, N. X., Hoanh, C. T. and Tuong, T. P.: Sea level rise affecting the Vietnamese Mekong Delta: water elevation in the flood season and implications for rice production, Clim. Change, 66(1–2), 89–107, doi:10.1023/B:CLIM.0000043144.69736.b7, 2004.

Weatherall, P., Marks, K. M., Jakobsson, M., Schmitt, T., Tani, S., Arndt, J. E., Rovere, M., Chayes, D., Ferrini, V. and Wigley, R.: A new digital bathymetric model of the world's oceans, Earth Sp. Sci., 2, 331–345, 2015.

Webster, R. and Oliver, M. A.: Geostatistics for Environmental Scientists, Second Edi., John Wiley & Sons, Ltd., 2007.

van der Wegen, M., Jaffe, B. E. and Roelvink, J. A.: Process-based, morphodynamic hindcast of decadal deposition patterns in San Pablo Bay, California, 1856-1887, J. Geophys. Res., 116, 1–22, doi:10.1029/2009JF001614, 2011.

Van der Wegen, M. and Roelvink, J. A.: Reproduction of estuarine bathymetry by means of a process-based model: Western Scheldt case study, the Netherlands, Geomorphology, 179, 152–167, doi:10.1016/j.geomorph.2012.08.007, 2012.

Wille, K.: Development of automated methods to improve surface modeling of river channel geometry and features., 2013.

Williams, G. P.: Sediment concentration versus water discharge during single hydrologic events in rivers, J. Hydrol., 111, 89–106, doi:10.1016/0022-1694(89)90254-0, 1989.

Willmott, C. J.: On the validation of models, Phys. Geogr., 2(2), 184–194, doi:10.1080/02723646.1981.10642213, 1981.

Winterwerp, J. C., Manning, A. J., Martens, C., de Mulder, T. and Vanlede, J.: A heuristic formula for turbulence-induced flocculation of cohesive sediment, Estuar. Coast. Shelf Sci., 68(1), 195–207, doi:10.1016/j.ecss.2006.02.003, 2006.

Wolanski, E., Huan, N. N., Dao, L. T., Nhan, N. H. and Thuy, N. N.: Fine-sediment dynamics in the Mekong River estuary, Viet Nam, Estuar. Coast. Shelf Sci., 43, 565–582, 1996.

Wolanski, E., Nhan, N. H. and Spagnol, S.: Sediment dynamics during low flow conditions in the Mekong River Estuary, Vietnam, J. Coast. Res., 14(2), 472–482, 1998.

Wong, P. P., Losada, I. J., Gattuso, J.-P., Hinkel, J., Khattabi, A., McInnes, K. L., Saito, Y. and Sallenger, A.: Coastal systems and low-lying areas, in Climate change 2014: Impacts, adaptation, and vulnerability. Part A: Global and sectoral aspects. Contribution of working group II to the fifth assessment report of the Intergovernmental Panel on Climate Change, edited by and L. L. W. Field, C.B., V.R. Barros, D.J. Dokken, K.J. Mach, M.D. Mastrandrea, T.E. Bilir, M. Chatterjee, K.L. Ebi, Y.O. Estrada, R.C. Genova, B. Girma, E.S. Kissel, A.N. Levy, S. MacCracken, P.R. Mastrandrea, pp. 361–409, Cambridge University Press, Cambridge, United Kingdom and New York, NY, USA., 2014.

Wood, M., Hostache, R., Neal, J., Wagener, T., Giustarini, L., Chini, M., Corato, G., Matgen, P. and Bates, P.: Calibration of channel depth and friction parameters in the LISFLOOD-FP hydraulic model using medium-resolution SAR data and identifiability techniques, Hydrol. Earth Syst. Sci., 20(12), 4983–4997, doi:10.5194/hess-20-4983-2016, 2016.

Wu, J., Liu, J. T. and Wang, X.: Sediment trapping of turbidity maxima in the Changjiang Estuary, Mar. Geol., 303–306, 14–25, doi:10.1016/j.margeo.2012.02.011, 2012.

Wu, Y., Falconer, R. A. and Struve, J.: Mathematical modelling of tidal currents in mangrove forests, Environ. Model. Softw., 16(1), 19–29, doi:10.1016/S1364-8152(00)00059-1, 2001.

Xing, F., Meselhe, E. A., Allison, M. A. and Weathers, H. D.: Analysis and numerical modeling of the flow and sand dynamics in the lower Song Hau channel, Mekong Delta, Cont. Shelf Res., 147(August), 62–77, doi:10.1016/j.csr.2017.08.003, 2017.

Xue, Z., Liu, J. P., DeMaster, D., Van Nguyen, L. and Ta, T. K. O.: Late Holocene evolution of the Mekong subaqueous Delta, Southern Vietnam, Mar. Geol., 269(1–2), 46–60, doi:10.1016/j.margeo.2009.12.005, 2010.

Xue, Z., He, R., Liu, J. P. and Warner, J. C.: Modeling transport and deposition of the Mekong River sediment, Cont. Shelf Res., 37, 66–78 [online] Available from: http://dx.doi.org/10.1016/j.csr.2012.02.010, 2012.

Yamazaki, D., Ikeshima, D., Tawatari, R., Yamaguchi, T., O'Loughlin, F., Neal, J. C., Sampson, C. C., Kanae, S. and Bates, P. D.: A high-accuracy map of global terrain elevations, Geophys. Res. Lett., 44, 5844–5853, doi:10.1002/2017GL072874, 2017.

Yu, W., Kim, Y., Lee, D. and Lee, G.: Hydrological assessment of basin development scenarios: Impacts on the Tonle Sap Lake in Cambodia, Quat. Int., (September), 0–1, doi:10.1016/j.quaint.2018.09.023, 2018.

Zhang, Y., Xian, C., Chen, H., Grieneisen, M. L., Liu, J. and Zhang, M.: Spatial interpolation of river channel topography using the shortest temporal distance, J. Hydrol., 542, 450–462, doi:10.1016/j.jhydrol.2016.09.022, 2016.

LIST OF ACRONYMS

1D	One Dimentional
2D	Two Dimensional
3D	Three Dimentional
SSC	Suspended sediment concentration
VMD	the Vietnamese Mekong Delta
CMD	the Cambodian Mekong Delta
CV border	the Vietnam-Cambodia border
AR	Anisotropy-removed interpolation
OK	Ordinary Kriging
KED	Kriging with External Drift
IDW	Inverse Distance Weighting
R	Coefficient of correlation
RMSE	Root mean square error
ME	Mean error
MRC	Mekong River Commision
LXQ	Long Xuyen Quadrangle
POR	Plain of Reeds
LMB	the Lower Mekong Basin

LIST OF TABLES

LIST OF FIGURES

ABOUT THE AUTHOR

Vo Quoc Thanh was born on July 12th 1986 in Bac Lieu, Vietnam. He completed his BSc degree in Environmental Engineering from Can Tho University, Vietnam in 2008. He obtained his MSc degree in Environmental Management at Can Tho University, Vietnam in 2012.

After graduated, he has worked as a researcher in Department of Environmental and Natural Resources Management, Can Tho University until 2013. He has involved in several projects on flood hazard mapping and water infrastructure. From 2013 to 2015, he has worked as a lecturer in Department of Environmental and Natural Resources Management, Can Tho University. During this period, he also acted as project assistant of the "Climate change and affecting land use in the Mekong Delta: Adaptation of rice-based cropping systems" project funded by ACIAR. From 2015 onwards he is a lecturer in Department of Water Resources, Can Tho University. From 2015 to 2019, he has involved in the project of "Modeling a sparsely-sampled, complex delta system: Mekong Delta case study", funded by the Office of Naval Research and this work was financed by this fund.

He is married to Linh and they have a daughter and a son.

Journals publications

Thanh, V. Q., Roelvink, D., van der Wegen, M., Reyns, J., van der Spek, A., Vinh, G.V. and Linh, V.T.P.: Suspended sediment dynamics and budget in the Mekong Delta: A numerical investigation, Cont. Shelf Res., 2020 (under review).

Thanh, V. Q., Roelvink, D., van der Wegen, M., Tu, L. X., Reyns, J. and Linh, V. T. P.: Spatial topographic interpolation for meandering channels, J. Waterw. Port, Coastal, Ocean Eng., 146(5), doi: 10.1061/(ASCE)WW.1943-5460.0000582, 2020.

Thanh, V. Q., Roelvink, D., van Der Wegen, M., Reyns, J., Kernkamp, H., Vinh, G. Van and Linh, V. T. P.: Flooding in the Mekong Delta: the impact of dyke systems on downstream hydrodynamics, Hydrol. Earth Syst. Sci., 24, 189–212, 2020.

Linh, V. T. P., Hoang, L. V., **Thanh, V. Q.**, 2019. Application of Landsat images to estimate suspended sediment concentration in the Hau and Tien rivers. Can Tho University Journal of Science. No 55(2): 134-144 (Vietnamese).

Tu, L. X., **Thanh, V. Q.**, Reyns, J., Van, S. P., Anh, D. T., Dang, T. D. and Roelvink, D.: Sediment transport and morphodynamical modeling on the estuaries and coastal zone of the Vietnamese Mekong Delta, Cont. Shelf Res., 186, 64–76, doi:10.1016/j.csr.2019.07.015, 2019.

Meselhe, E., Roelvink, D., Wackerman, C., Xing, F. and **Thanh, V. Q.**: Modeling the process response of coastal and deltaic systems to human and global changes: Focus on the Mekong system, Oceanography, 30(3), doi:10.5670/oceanog.2017.317, 2017.

Thanh, V. Q., Reyns, J., Wackerman, C., Eidam, E. F. and Roelvink, D.: Modeling suspended sediment dynamics on the subaqueous delta of the Mekong River, Cont. Shelf Res., 147(August), 213–230, doi:10.1016/j.csr.2017.07.013, 2017.

Conferences

Thanh, V. Q., Roelvink, D., van der Wegen, M., and Reyns, J.: Numerical modeling of suspended sediment dynamics in the Mekong Delta. Coastal Engineering Proceedings, (36v), sediment.24. https://doi.org/10.9753/icce.v36v.sediment.24, 2020.

Manh-Hung Le, Venkataraman Lakshmi, **Thanh, V. Q.**, Hong Minh Hoang, Tri Pham Dang Van: Assessment of hydrological processes in a polder of Mekong Delta using SWAT+ model, American Geophysical Union, Fall Meeting 2019, abstract #H11I-1580, 2019.

Thanh, V. Q., Reyns, J., Wackerman, C., Eidam, E. F. and Roelvink, D.: Modeling suspended sediment dynamics on the subaqueous delta of the Mekong River, IHE Delft PhD Symposium 2017: Climate extremes and water management challenges, 2017.

Thanh, V. Q., Reyns, J., Kernkamp, H., Roelvink, J.A., Van der Wegen, M.: Numerical Modeling of Tidal Dynamics and Transport in the Multi-channel Estuary of the Mekong River, American Geophysical Union, Ocean Sciences Meeting 2016, abstract #MG54B-2030, 2016.

Roelvink, J.A., Reyns, J., McLachlan, R.L., Eidam, E.F., Liu, P., Ogston, A.S., **Thanh, V. Q.**, Wackerman, C.: Integrating 3D Modeling, In-Situ and Remote-Sensed Observations of Flow and Sediment Dynamics in the Hau River Estuary and Shelf, Mekong Delta, Vietnam, American Geophysical Union, Ocean Sciences Meeting 2016, abstract #MG51A-08, 2016.

Netherlands Research School for the
Socio-Economic and Natural Sciences of the Environment

D I P L O M A

for specialised PhD training

The Netherlands research school for the
Socio-Economic and Natural Sciences of the Environment
(SENSE) declares that

Vo Quoc Thanh

born on 12 July 1986 in Bac Lieu, Vietnam

has successfully fulfilled all requirements of the
educational PhD programme of SENSE.

Delft, 13 April 2021

Chair of the SENSE board

Prof. dr. Martin Wassen

The SENSE Director

Prof. Philipp Pattberg

The SENSE Research School has been accredited by the Royal Netherlands Academy of Arts and Sciences (KNAW)

K O N I N K L I J K E   N E D E R L A N D S E
A K A D E M I E   V A N   W E T E N S C H A P P E N

The SENSE Research School declares that Vo Quoc Thanh has successfully fulfilled all requirements of the Educational PhD Programme of SENSE with a work load of 42.1 EC, including the following activities:

SENSE PhD Courses

o   Environmental research in context (2015)
o   SENSE writing week (2016)
o   Research in context activity: 'Coastal Engineering and Port Development Meetings' (2019)

Other PhD and Advanced MSc Courses

o   Becoming a creative researcher, TU Delft (2016)
o   Geostatistics for water management and environmental sciences, IHE Delft (2017)
o   Where there is little data: How to estimate design variables in poorly gauged basins, IHE Delft (2017)
o   NCK Summer School: Estuarine and coastal processes in relation to coastal zone management, Netherlands Centre for Coastal Research (2017)
o   Speedreading and mindmapping, TU Delft (2017)
o   Effective management of your PhD research (2017)
o   Discovering statistics using SPSS (2017)
o   Achieving your goals and performing more successfully in your PhD (2017)
o   Data visualisations: A practical approach (2017)
o   Creative Tools for Scientific Writing (2017)
o   Career Development- Preparing for your next career step in academia (2017)

Oral and Poster Presentations

o   *Interactions between hydrodynamics and morphology dynamics in alluvial estuaries*, PhD symposium, 28-29 September 2015, Delft, The Netherlands
o   *Numerical modeling of tidal dynamics and transport in the multi-channel estuary of the Mekong River*, Ocean Sciences Meeting, Feb 21-26th 2016, New Orleans, United States of America
o   *Modelling suspended sediment dynamics on the subaqueous delta of the Mekong River.* IHE PhD symposium, 2-3 September 2017, Delft, The Netherlands

SENSE coordinator PhD education

Dr. ir. Peter Vermeulen